Joachim Franck

Systematic Study of Key Components for a Coaxial-Cavity Gyrotron for DEMO

Karlsruher Forschungsberichte aus dem
Institut für Hochleistungsimpuls- und Mikrowellentechnik

Herausgeber: Prof. Dr.-Ing. John Jelonnek

Band 11

Systematic Study of Key Components for a Coaxial-Cavity Gyrotron for DEMO

by
Joachim Franck

KIT Scientific Publishing

Dissertation, Karlsruher Institut für Technologie (KIT)
Fakultät für Elektrotechnik und Informationstechnik, 2017
Referenten: Prof. Dr.-Ing. John Jelonnek
Prof. Dr. Olgierd Dumbrajs

Impressum

KIT Scientific Publishing

Karlsruher Institut für Technologie (KIT)
KIT Scientific Publishing
Straße am Forum 2
D-76131 Karlsruhe

www.ksp.kit.edu

Print on Demand 2017 – Gedruckt auf FSC-zertifiziertem Papier
ISSN 2192-2764
ISBN 978-3-7315-0652-2
DOI: 10.5445/KSP/1000068000

Foreword of the Editor

Gyrotrons are vacuum tubes based on the electron cyclotron maser instability. They are the only microwave sources that are capable of efficiently generating continuous-wave (CW) power at megawatt levels. As such, gyrotron oscillators will be used for electron cyclotron heating, current drive, and stabilization of the magnetically confined plasma in future nuclear fusion power plants, including the first demonstration power plant (DEMO).

Already today, gyrotrons are used for the same purpose in plasma experiments such as ASDEX-Upgrade and Wendelstein 7-X, and they will be used in the ITER experiment accordingly. These state-of-the-art gyrotrons deliver up to 1 MW CW output power at single frequencies of up to 170 GHz. However, gyrotrons for DEMO shall deliver an output power of 2 MW at frequencies of up to 240 GHz. Additionally, possible operation at multiples of the RF window's resonance frequency is required. Nonetheless, today's technical limitations for emitter current density and thermal loading of the cavity wall still have to be considered. The latter implies a highly oversized gyrotron cavity, hence operation at very high order modes. This, in turn, requires methods to reduce mode competition. One method is to use a coaxial insert that extends from the magnetron injection gun (MIG) into the cavity; its introduction also reduces space-charge effects of the electron beam. All these conditions combined question the possibility to obtain a realistic design. A design process that answers this would encompass systematic selection of the operating modes and proper design of cavity (including coaxial insert) and MIG, followed by tolerance studies with regards to electron beam misalignment and to the coaxial insert.

In his work, Dr.-Ing. Joachim Franck proposes the physical design of the two key components – MIG and cavity – for a coaxial-cavity 2 MW CW gyrotron with a main operating frequency of 238 GHz and auxiliary frequencies at 204 GHz and 170 GHz. The design is based on present-day technical boundary conditions. He derives and implements a holistic design process and selects the operating modes according to a novel systematic multi-frequency mode selection approach. The advanced triode-MIG he develops corresponds to a realistic 10.5 T magnet design, obtained in the frame of this work. Multi-frequency behavior of MIG and cavity is investigated in detail. Dr.-Ing. Joachim Franck analyses the most severe effects of a too small number of corrugations and of possible insert misalignment: Mode deterioration and voltage depression variation. The study of mode deterioration due to insert misalignment is based on an expansion of the electromagnetic field into eigenmodes, and the effect of voltage depression due to insert misalignment is simulated using an efficient method.

Dr.-Ing. Joachim Franck provides the gyrotron community with a solid starting point for more detailed design activities for high-power coaxial-cavity gyrotrons. The general design concepts and tools he developed are very valuable for future studies to maximize the possible gyrotron performance.

Systematic Study of Key Components for a Coaxial-Cavity Gyrotron for DEMO

Zur Erlangung des akademischen Grades eines
Doktor-Ingenieurs
von der Fakultät für
Elektrotechnik und Informationstechnik
des Karlsruher Instituts für Technologie (KIT)

genehmigte
Dissertation
von

Dipl.-Phys. Joachim Franck
geb. in Potsdam

Tag der mündlichen Prüfung:	21. Februar 2017
Hauptreferent:	Prof. Dr.-Ing. John Jelonnek
Korreferent:	Prof. Dr. Olgierd Dumbrajs

Kurzfassung

In der vorliegenden Arbeit wird das physikalische Design der Kernkomponenten – Wechselwirkungsraum und Elektronenquelle (MIG) – eines koaxialen 238-GHz-, 2-MW-Dauerstrichgyrotrons, welches für zukünftige Fusionskraftwerke geeignet ist, erarbeitet. Für beide Komponenten wird das Mehrfrequenzverhalten untersucht. Die Arbeit beinhaltet neue systematische Ansätze für Modenauswahl, Magnetspezifikation und MIG-Design um die hohen Frequenz- und Leistungsanforderungen, die an das Gyrotron gestellt werden, zu erfüllen. Obwohl eine ambitionierte Arbeitsmode sehr hoher Ordnung untersucht wird, werden die Grenzen des heute technisch Machbaren während des Designs berücksichtigt.

Gemeinsam mit dem Vorteil der höheren möglichen Ausgangsleistung haben koaxiale Gyrotrons den Nachteil, dass der Innenleiter gegenüber der Resonatorwand und/oder dem Elektronenstrahl versetzt sein kann, was die Betriebsstabilität beeinträchtigt. Während einige diesbezügliche Aspekte in der Vergangenheit untersucht wurden, präsentiert diese Arbeit zum ersten Mal eine robuste, systematische Analyse der kritischsten Effekte des Innenleiterversatzes, nämlich Störung der Moden und Variation der Spannungsabsenkung. Weiterhin wird eine Analyse der Abmessungen des Innenleiters und der Anzahl seiner Korrugationen durchgeführt.

Die vorhandene Theorie der Modendeformation wurde überarbeitet und erweitert, um den Versatz des Innenleiters noch besser behandeln zu können. Im Zuge dessen wurde eine systematische, präzise Schreibweise erarbeitet.

Abstract

In this work, the physical design of key components – cavity and magnetron injection gun (MIG) – for a coaxial-cavity 238 GHz 2 MW CW gyrotron, being suitable for future fusion power plants, is developed. Multi-frequency behavior is investigated for both components. This work includes new systematic approaches towards mode selection, magnet requirements, and MIG design in order to fulfill the challenging requirements on the gyrotron concerning high frequency and power. Although an ambitious very-high-order operating mode is considered, present-day technological boundaries are respected throughout the design.

Alongside with their advantage to allow higher output power, coaxial-cavity gyrotrons have the disadvantage of possible insert misalignment versus cavity wall and/or versus electron beam, which decreases operation stability. While several aspects of this have been investigated in the past a robust, systematic analysis of the most severe effects of insert misalignment – mode deterioration and voltage depression variation – is presented in this work for the first time. An analysis of insert dimensions and its number of corrugations has also been made.

In order to treat insert misalignment more properly, the existing theory of mode deformation was revised and extended. In the course of this a systematic, concise notation was developed.

Contents

Acronyms, Abbreviations, and Symbols

Acronym / Abbreviation	Full term
ACI	After-Cavity Interaction
BCI	Before-Cavity Interaction
BWO	Backward-Wave Oscillator
CARM	Cyclotron Autoresonance Maser
CC	Compensation Coil (a.k.a. Bucking Coil)
CVD	Chemical Vapor Deposition
CW	Continuous Wave (synonymous to *long-pulse* as opposed to SP: pulses longer than one second)
DEMO	Demonstration Power Plant
DNP-NMR	Dynamic Nuclear Polarization in Nuclear Magnetic Resonance (Spectroscopy)
ECCD	Electron Cyclotron Current Drive
EC	Electron Cyclotron (Resonance Heating)
ECM	Electron Cyclotron Maser
EGYC	European Gyrotron Consortium
FEL, FEM	Free Electron Laser; Free Electron Maser
FEM	Finite Element Method
FZK, FZKA	Forschungszentrum Karlsruhe (Research Center Karlsruhe, now part of KIT)
GC	Gun Coil
HD	High Depression
IC	Ion Cyclotron (Heating)
IHM	Institut für Hochleistungsimpuls- und Mikrowellentechnik (Institute for Pulsed Power and Microwave Technology)

Acronym / Abbreviation	Full term
ITER	Latin *iter* "route", "journey", "the way"; initially: International Thermonuclear Reactor
JAEA	Japan Atomic Energy Agency
KIT	Karlsruhe Institute of Technology
LD	Low Depression
LH	Lower Hybrid (Heating)
LOG	Large-Orbit Gyrotron
LTS	Low-Temperature Superconductor
MC	Main Coil
MIG	Magnetron Injection Gun
MHD	Magnetohydrodynamics
MSDC	Multi-Stage Depressed Collector
NBI	Neutral Beam Injection
NTM	Neoclassical Tearing Mode
PFC	Poloidal Field Coil
QOL	Quasi-Optical Launcher
RF	Radio Frequency (denoting any oscillating electro-magnetic field)
RMS	Root Mean Square
SCM	Superconducting Magnet
SDC	Single-stage Depressed Collector
SHM	Space Harmonic Method
SIE	Singular Integral Equation
SIM	Surface Impedance Model
SP	Short Pulse (as opposed to CW: pulses shorter than a few milliseconds)
TFC	Toroidal Field Coil
TLK	Tritium Laboratory Karlsruhe
TWT	Travelling Wave Tube
UH	Upper Hybrid (Heating)
VLD	Very Low Depression

The following are symbols used in this document as well as in the codes *Design-o-mat*, *WickedQueen*, and *SCNCHIMP*. In general, capital letters denote fixed or controllable/input values (e.g. geometry) while small letters denote reacting values (e.g. beam parameters). As an example, the maximum axial distance between emitter and cavity would be denoted $L_{\mathrm{EO,max}}$ and `l1e1o1mx`. For numerical values of physical quantities, see [O+14].

Symbol	Code	Description of variable	Usage as index
A	`a1`	Electromagnetic (four-)vector potential	Main anode (body)
a	`a3`	Taper factor $\tan(\Theta)$; numerical coefficient	
α_0	`a5`	Fine-structure constant $\approx 1/137.036$	
α	`a6`	Pitch or Brillouin angle, $\arctan(\beta_1/\beta_2)$; angular bipolar coordinate	
B	`b1`	Magnetic flux density (magnetic field)	B-field, magnetic ...
b	`b3`	Misalignment parameter (bipolar coordinates)	Beam(let), guiding center (center of gyration)
β	`b6`	(Electron) normalized velocity v/c	
C	`c1`	Dimensionless ratio of two values (e.g. magnetic compression, relative caustic radius)	Coordinate relative to guiding center (coordinate of cyclotron motion)
c	`c3`	Speed of light (299.792458 mm/ns)	
Ξ	`c4`	Arbitrary term	
ξ	`c6`	Axial variable (equivalent to z-coordinate or time)	

Symbol	Code	Description of variable	Usage as index
D	d1	Distance (depth D_r, period D_φ, thickness); electric displacement field	Design
d	d2	Differential	
d	d3	Corrugation parameter	(Voltage) depression
Δ	d4	(Maximum) step of ... (in numerical calculations)	
δ	d6	Difference, deviation, variation, detuning; error; Dirac's Delta function, Kronecker's Delta	
E	e1	Electric field	MIG: emitter (cathode, central)
e	e2	Elementary charge ($\approx 1.602177 \cdot 10^{-19}$ C)	Electron (or particle in general)
e	e3	Base of natural logarithm	
ε	e5	Electric permittivity (vacuum: $\varepsilon_0 = 1/(\mu_0 c^2)$)	
ϵ	e6	Levi-Civita symbol; unit vector along coordinate axis	
F	f1	Electromagnetic field tensor (E-B)	
f	f3	(Ordinary) frequency	
Φ	f4	Scalar electromagnetic potential	
ϕ	f5	Slow (gyro-)phase	
φ	f6	Azimuth (cylinder or polar coordinates)	Azimuthal component
G	g1	Macroscopic field tensor (D-H); coupling factor	MIG (as a whole), MIG region
g	g3	Pitch factor $\tan(\alpha)$	
Γ	g4	Axial electric field profile (arbitrary normalization)	
γ	g6	Relativistic factor $1/\sqrt{1 - v^2/c^2}$	Photons: Brillouin angle/length

Symbol	Code	Description of variable	Usage as index
H	h1	Height (length along z-axis, see L); magnetizing field	Magnet
$\hbar$	h2	Reduced Planck's constant ($\approx 1.054572 \cdot 10^{-34}$ J s)	
h	h3	Mode character (whispering-gallery vs. volume mode); mapping Cartesian – bipolar coordinates	
η	h6	Efficiency	
I	i1	Electric current	Inner surface of cavity (i.e. coaxial insert)
i	i2	Imaginary unit $\sqrt{-1}$	
i	i3	Counter (usually 1 ... n)	Spacial index 1
ι	i6	Particle current density	
J	j1	Bessel function	Coaxial MIG: inner parts; Bessel coefficient
$\jmath$	j2	Four-dimensional charge density	
j	j3	Electric current density	Counting variable (for component sections); spacial index 2
K	k1	Characteristic function/matrix	Caustic
k	k3	Wavenumber $2\pi/\lambda$	Spacial index 3
κ	k6	One-dimensional charge density	
L	l1	Length parallel (or projected) to z-axis; Bessel or Neumann function	Launcher
l	l3	Axial (longitudinal) index; distance along path	Axial (longitudinal) index

Symbol	Code	Description of variable	Usage as index
Λ	l4	Lagrange density	
λ	l6	Wavelength	
M	m1	(Groove) number; mode index in rectangular waveguide; membrane function	Modulation (or inner) anode
$\mathrm{m_e}$	m2	Electron rest mass ($\approx$ $9.109383 \cdot 10^{-31}$ kg)	
m	m3	Azimuthal mode index; Bessel/Neumann order	Azimuthal mode index
μ_0	m5	Vacuum magnetic permeability ($4\pi \cdot 10^{-7}\ \mathrm{N/A^2}$)	
μ	m6	Dimensionless cavity length; inclination; permeability	Spacetime index 1
N	n1	Neumann function; mode index in rectangular wave-guide	MIG: nose; Neumann coefficient
n	n3	Total number of ...; refractive index; invariant density (number per volume)	End of counting (component sections)
ν	n6	Eigenvalue of (pure) inner mode	Spacetime index 2
O	o1		(Outer surface of) cavity
P	p1	(Radiation) power; location of poles in a bipolar coordinate system	MIG: prolongator; plasma
p	p3	Radial mode index; root index	Radial mode index
ϖ	p5	Corrugation term as defined in Eq. B.19	
π	p6	Pi (3.14159...); momentum density	
Q	q1	Quality factor; general electric charge	Probe
$\mathrm{q_e}$	q2	Electric charge of particle	

Symbol	Code	Description of variable	Usage as index
q	q3	Substitute azimuthal index	Substitute azimuthal index
Ψ	q4	Magnetic flux	
ψ	q6	Spread angle (azimuthal direction)	
R	r1	Defined length in radial direction, measured from z-axis	(Whole) cavity (resonator)
r	r3	Radius coordinate (cylindrical coordinate); substitute azimuthal index	Radial component; substitute azimuthal index
ρ	r5	Charge density; radial bipolar coordinate	
ϱ	r6	Scalar component of four-dimensional density	
S	s1	Surface; Poynting vector	Skin
s	s3	Cyclotron harmonic	Cyclotron harmonic
Σ	s4	Action; surface charge density (charge per area)	
ς	s5	Absolute spread of ... (depending upon definition)	
σ	s6	Electric conductivity	Spacetime index 3
T	t1	Temperature	Beam tunnel
TE	—	Transverse electric (H mode)	—
TM	—	Transverse magnetic (E mode)	—
TEM	—	Transverse electromagnetic (S mode)	—
t	t3	Time (coordinate)	Time component
τ	t6	Time-interval / lifetime; proper time	

Symbol	Code	Description of variable	Usage as index
U	u1	Voltage (electric potential difference)	Uptaper
u	u3	Dimensionless momentum	
Ω	u4	Angular frequency (rotating particle)	Ohmic
ω	u6	Angular frequency (electromagnetic wave; plasma)	
V	v1	Volume	
v	v3	Phase/group velocity (particle speed: β, u, y)	
Θ	v4	Polar angle (component surface or static field)	
θ	v5	Polar angle coordinate	Polar component
W	w1	Energy (e.g. of electron)	Output window(s)
w	w3	Energy flux (energy per time and area; power per area)	Generalized or combined mode index $w = (m, p)$
X	x1	Placeholder variable	
x	x3	Placeholder variable; 1st Cartesian coordinate	
χ	x6	Eigenfunction argument	
Y	y1	(Alternative notation for Neumann functions)	Collector
y	y3	Relativistic four-velocity; 2nd Cartesian coordinate	
Υ	y4	Arbitrary term	
Z	z1	Impedance	
Z_0	z2	Impedance of free space ($\approx 120\pi\ \Omega$)	
z	z3	Axial/longitudinal (3rd Cartesian) coordinate	Axial component
ζ	z6	Relative spread ς_X / X; mode-series rating	

Symbol	Code	Description of variable	Usage as index
appr	ap	—	Approximate value
co	co	—	Co-rotating
ct	ct	—	Counterrotating
diff	df	—	Diffractive
dom	dm	—	Dominant value
dr	dr	—	Drift
encl	ec	—	Enclosed
end	ed	—	End/final value
elec	el	—	Electronic
gr	gr	—	Group
in	in	—	Input value; inner coordinate system
min	mn	—	Minimum/lower end value
$\underline{x}$	mt	—	(2D) matrix
max	mx	—	Maximum/upper end value
new	nw	—	New value
old	od	—	Old value
opt	op	—	Optimum value
out	ot	—	Output value; outer coordinate system
ph	ph	—	Phase
$\hat{x}$	pk	—	Peak value, amplitude

Symbol	Code	Description of variable	Usage as index
$\parallel$	`pl`	—	Parallel to z-axis
$\perp$	`pp`	—	Perpendicular to z-axis (might be expressed as complex value)
rel	`rl`	—	Relative
start	`st`	—	Start/initial value
tang	`tg`	—	Tangential
typ	`tp`	—	Typical value
$\vec{x}$	`vc`	—	Vector (1D matrix)
∂	—	Partial derivative ($\partial_x \equiv \frac{\partial}{\partial x}$); boundary of ...	—
∇	—	Nabla operator	—
Δ	—	Laplace operator	—
$\square$	—	D'Alembert operator	—

Introduction and Motivation

Nuclear fusion promises to provide reliable energy from abundant resources while causing much less dangerous radioactive waste than fission power plants. In magnetic confinement fusion, the fuel plasma has to be heated to above 100 million kelvins, which can be achieved via electron cyclotron heating with microwaves produced by gyrotrons, among other heating techniques. In order to meet the requirements for the first demonstration fusion power plants, these gyrotrons have to deliver more than 1 MW in stable, continuous-wave (CW) operation with high efficiency at frequencies above 200 GHz. These values are beyond present practical experience, mainly due to the very high eigenvalue of the required operating mode. For usage of the same basic tube design for various tasks or in various facilities, operation at several distant frequencies, termed multi-frequency operation, is worth being investigated.

In this work, a realistic physical design for cavity and magnetron injection gun (MIG) of a 2 MW CW gyrotron is developed, with main operating frequency of 237.5 GHz, suitable for a DEMO reactor, and auxiliary frequencies at 170.0 GHz (ITER) and 203.8 GHz. A general, systematic multi-frequency mode-selection scheme was developed for this purpose, and a realistic magnet design was obtained and used as a basis for the MIG. All designs respect present-day technological restrictions.

The design itself is based on the coaxial-cavity concept, which promises high output power and stability at very-high-order modes. However, the coaxial insert requires careful study of its dimensions, surface corrugations, and effects of possible misalignment versus electron beam and/or cavity wall, which has been done based on a new method for voltage depression calculation in misaligned cavities and on an extension of the existing theory of modes in misaligned coaxial cavities. The phenomenon of inner modes, which is a characteristic feature of gyrotrons with corrugated insert, has been investigated in detail.

In order to achieve a consistent notation for variables, a systematic notation throughout this work and related computer codes was developed. Existing formulas and derivations were revised and in some cases formulated in a more elegant way, improving their readability, especially when translated into computer code.

1 Gyrotrons in the Context of Nuclear Fusion

1.1 Nuclear Fusion

1.1.1 General Considerations

The global demand for energy is expected to grow by a factor of six by the end of this century [LCV$^+$05] due to both the growing world population and an increasing average power consumption per capita, which in the past has always been connected with prosperity [GS08]. While fossile resources (oil, coal, and gas) are limited on Earth and contribute significantly to the climate change [GS08, CSB$^+$13], renewable sources are strongly dependent on weather conditions (wind), require space and depend on special environmental conditions (solar) or have a negative impact on our ecosystem (hydro), fauna (wind), or on food production (bio fuels). Thus, it is undoubtly important to investigate continuously available energy sources in addition to the mentioned renewables, especially since energy diversification also poses a strategic advantage.

Aside solar power, controlled nuclear fission and nuclear fusion are considered the most yielding energy sources accessible on Earth [IPP03]. Even if one assumes perfectly safe fission power plants, about $1000\ \mathrm{m}^3$ of nuclear waste are produced per generated GWh of electricity, of which 7 % are high-level waste, i.e. waste with thermal power above $2\ \mathrm{kW}/\mathrm{m}^3$ and/or half-life longer than 200000 years [Dat13]. Nuclear waste poses a latent danger to the ecosystem – over timespans which are much beyond our capabilities of effective planning[1] – since it can cause cancer and genetic defects in living beings. At present there are no practicable, large-scale technologies to fully protect the ecosystem from such waste, such as removal into space or partitioning and transmutation (P&T) to non-radioactive nuclei. So far, the best strategy to minimize the danger from radioactive waste is deep geological disposal [AM07]. However, site selection and confinement materials characterization for geological disposal are always based on short-term considerations, such as current technology, model assumptions, and political circumstances.

Nuclear fusion power plants, in contrary, could provide reliable energy from abundant resources. The only genuinely radioactive material associated with fusion is the hydrogen isotope tritium [Tos13], one of the fuels. The half-life of tritium is only 12.3 years, it decays into non-radioactive particles and it emits very low-energy beta radiation (5.69 keV on average) with a range of 6 mm in air and 6 μm in water. This makes tritium hazardous only if it is taken inside the body, mostly via inhalation and ingestion of tritiated water. However, it cannot be stored in living organisms permanently: its average biological half-life in humans is 9.7 days. Most importantly, it is not at all foreseen for the tritium to leave the power plant, since it would be needed as fuel within the facility. Safe tritium handling in a closed loop has thus been an important topic in fusion reactor design from the beginning. Fusion power plants as a whole are inherently safe because the stored amount of tritium would be small and no nuclear chain reaction can take place [EC09].

The second class of radioactive materials associated with nuclear fusion are materials which are activated by the neutron flux leaving the fusion plasma [KGOP90]. Already in the 1980s studies have been conducted to

[1] Insulation from the environment has to be maintained for around 10 to 20 half-lives [FHKS89]. Half-lives for the most siginifcant radioactive waste are 24400 years for Pu-239; 210000 years for Tc-99; 380000 years for Pu-242; 2.1 million years for Np-237; 15.7 million years for I-129.

find suitable reduced-activation alloys for use in fusion reactors. Still today it is consensus that only such low-activation materials may be used for plasma-facing components in fusion reactors, and the search for and qualification of such materials is one, maybe the biggest, obstacle in current reactor design activities. This approach leads to heterogeneous, but almost entirely [EG06, PEK$^+$12] low-level activated waste with relatively short half-lifes. In this respect, it is comparable to the waste from high-energy particle experiments or from medical applications [AM07].

According to our present understanding the constituents of atomic nuclei, protons and neutrons, consist of quarks bound together by the strong force, one of the four known fundamental interactions alongside electromagnetism, weak interaction, and gravitation. In fact, the binding energy itself makes up over 98 % of the nucleons' rest mass. In stable nuclei protons and neutrons are bound by a residual of their intra-nucleon interaction: the nuclear force. Still, this residual inter-nucleon force is not only strong enough that it holds together the positively charged protons, it also prevents the neutrons from decay [MK94]. The typical binding energy of the nuclear force is around 8 MeV per nucleon, which is a mass defect of about 1 % of the nucleons' weight [O$^+$14]. In former times, the nuclear force was seen as a fundamental force, with pions – instead of gluons – as force carriers. While this is only an approximation, the pion mass of around 140 MeV corresponds to an interaction length of the nuclear force of around 1.4 fm, which is quite exactly the nucleon radius.

A detailed plot of the binding energy per nucleon versus atomic mass number is given in Fig. 1.1. One can see that there is a clear maximum binding energy per nucleon for nuclei with around 60 protons and neutrons. Heavier nuclei can release a part of their energy through decay or fission, while lighter nuclei must fuse to convert binding energy to other forms of energy, predominantly kinetic energy. In fact, almost all nuclei heavier than hydrogen and helium are thought to have been produced in stars by nuclear fusion (stellar and supernova nucleosynthesis). One can also see that the left side of the curve is much steeper than the right part, which indicates that the released binding energy per involved nucleon is much larger for fusion than for fission reactions.

Some important nuclear reactions, including released (>0) or required (<0) binding energy, are the following [IPP03, MK94]:

$$\mathrm{p}+\mathrm{p} \rightarrow \mathrm{D}+\mathrm{e}^{+}+\nu \qquad +0.42\ \mathrm{MeV} \tag{1.1}$$

$$\mathrm{p}+\mathrm{D} \rightarrow {}^{3}\mathrm{He}+\gamma \qquad +5.49\ \mathrm{MeV} \tag{1.2}$$

$${}^{3}\mathrm{He}+{}^{3}\mathrm{He} \rightarrow {}^{4}\mathrm{He}+\mathrm{p}+\mathrm{p} \qquad +12.86\ \mathrm{MeV} \tag{1.3}$$

$$\mathrm{D}+\mathrm{D} \rightarrow {}^{3}\mathrm{He}+\mathrm{n} \qquad +3.27\ \mathrm{MeV} \tag{1.4}$$

$$\mathrm{D}+\mathrm{D} \rightarrow \mathrm{T}+\mathrm{p} \qquad +4.03\ \mathrm{MeV} \tag{1.5}$$

$$\mathrm{D}+\mathrm{T} \rightarrow {}^{4}\mathrm{He}+\mathrm{n} \qquad +17.58\ \mathrm{MeV} \tag{1.6}$$

$$\mathrm{D}+{}^{3}\mathrm{He} \rightarrow {}^{4}\mathrm{He}+\mathrm{p} \qquad +18.35\ \mathrm{MeV} \tag{1.7}$$

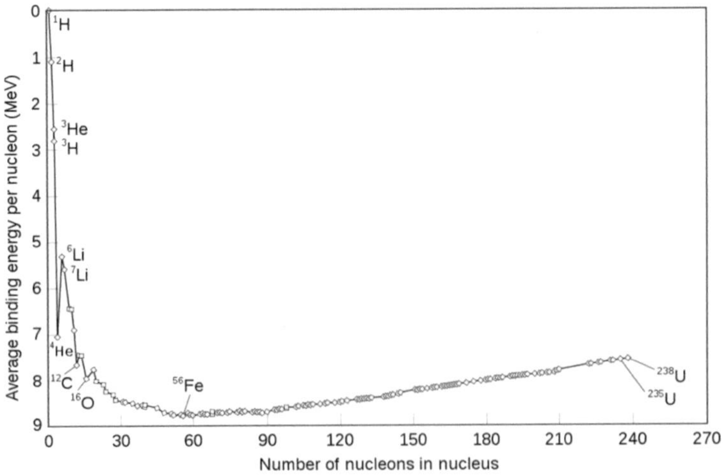

Figure 1.1: Nuclear binding energy per nucleon versus nucleon number of common nuclei. The vertical axis points downwards to indicate that increasing binding energy corresponds to increasing stability. See e.g. [IPP03, Sta05].

$$\mathrm{T} + \mathrm{T} \rightarrow {}^{4}\mathrm{He} + 2\,\mathrm{n} \qquad +11.3\ \mathrm{MeV} \tag{1.8}$$

$$\mathrm{p} + {}^{11}\mathrm{B} \rightarrow {}^{4}\mathrm{He} + {}^{4}\mathrm{He} + {}^{4}\mathrm{He} \qquad +8.7\ \mathrm{MeV} \tag{1.9}$$

$$\mathrm{n} + {}^{7}\mathrm{Li} \rightarrow {}^{4}\mathrm{He} + \mathrm{T} + \mathrm{n} \qquad -2.47\ \mathrm{MeV} \tag{1.10}$$

$$\mathrm{n} + {}^{6}\mathrm{Li} \rightarrow {}^{4}\mathrm{He} + \mathrm{T} \qquad +4.78\ \mathrm{MeV} \tag{1.11}$$

$$\mathrm{n} + {}^{235}\mathrm{U} \rightarrow \langle \text{heavy nuclei} + \text{neutrons} \rangle \qquad +200\ \mathrm{MeV} \tag{1.12}$$

$$\mathrm{n} + {}^{239}\mathrm{Pu} \rightarrow \langle \text{heavy nuclei} + \text{neutrons} \rangle \qquad +200\ \mathrm{MeV} \tag{1.13}$$

Hydrogen (protons (p) or protium (H), and deuterium (D)), helium-4 (^{4}He) and lithium (Li) are abundant and rather uniformly distributed on Earth, while helium-3 (^{3}He), tritium (T), uranium (U) and plutonium (Pu) are rare; besides this, the latter three are radioactive. Reactions 1.1 to 1.3 are the reactions through which hydrogen is fused to helium in stars (alongside the CNO cycle). Reaction 1.1 involves the weak interaction and has therefore a very small cross-section (reaction probability) [MK94]. Reactions 1.4 to 1.9 are those reactions of light nuclei which have a high reaction probability at accessible conditions, with 1.6 being the most probable and reaction 1.9 including neither radioactive isotopes nor fast nucleons as products. Reactions 1.10 and 1.11 produce tritium (breeding reactions) that can be used for reaction 1.6, involving fast and slow (thermal) neutrons, respectively [Sta10]. Reactions 1.12 and 1.13 are typical reactions in fission power plants, for comparison. From these considerations one has chosen reactions 1.6 and 1.10/1.11 for energy production and tritium breeding, respectively, in fusion power plants. Fig. 1.2 illustrates the two resulting fusion-breeding reaction pairs. One can see that the resulting fuel necessary for such a plant is only hydrogen and lithium, while only the noble gas helium is emitted. This simplified picture does not take into account materials activated by the involved neutrons, but the radiotoxicity of such materials can rather be compared with that from coal plants than that from conventional nuclear plants [GS08]. The energy yield per mass from deuterium-tritium fusion is about one million times higher than the yield from burned coal [GS08].

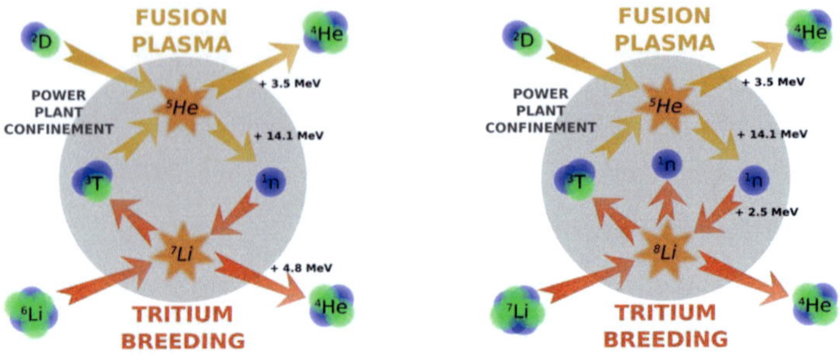

Figure 1.2: Reaction pairs in a D-T fusion power plant with lithium-6 (left) and lithium-7 (right) as sources for tritium breeding.

Beyond the nucleus radius the nuclear force drops quickly and the Coulomb repulsion between protons dominates. In order to fuse nuclei, this electromagnetic force has to be overcome before the binding energy of the nuclei can be accessed. It is therefore easier to fuse nuclei with small proton numbers: for the hydrogen isotopes one would only need a few hundred keV to overcome Coulomb repulsion directly. If the quantum-mechanical tunnelling effect is taken into account, this value drops to around 20 keV thermal energy (around $200 \cdot 10^6$ K) [IPP03]. At these energies, atoms or even molecules cannot be formed, so the nuclei and their unbound electrons are in the (macroscopically neutral) plasma state.

A plasma can retain its temperature if at least some of its constituents interact exothermally ("burn"), e.g. fuse, and if at least one reaction product transfers its kinetic energy to the other constituents. A rather simple energy balance leads to a characteristic fusion parameter (Triple Product), which for a D-T plasma – with alpha particles as the heating, high-energy product and with neutrons as the accessible source of kinetic energy – has to fulfill the Lawson criterion

$$n\,T\,\tau > 3 \cdot 10^{21}\ \mathrm{keV\ s\ m^{-3}} \tag{1.14}$$

where n is the number density of ions in m^{-3} (of both species; and equal to the number density of electrons), T is the temperature[2] in keV and τ is the energy confinement time in s [Sta05]. The electrically neutral neutrons leave the plasma and heat the surrounding confinement.

Without analyzing these numbers in detail, it is immediately clear that the conditions for nuclear fusion cannot be achieved easily on Earth, otherwise the existing fuel would have already burned up, e.g. due to lightning or meteorite impacts. One of the reasons for this is that the reactants for the most probable reactions – mainly deuterium, tritium and helium-3 – are rare or sparsely distributed on Earth. While there did exist natural fission reactors involving uranium-235 (Oklo/Gabon [MK94]), tritium has a 57 million times smaller half-life and can therefore not accumulate naturally to a sufficient degree. The other, more important reason is condition 1.14, which implies also a high temperature or a long confinement time to be given. There are two main technological concepts allowing to reach this condition in a controlled environment: inertial confinement fusion and magnetic confinement fusion.

Inertial Confinement Fusion

Deuterium and tritium are compressed in a target to typical solid-state densities of $10^{27}\ \mathrm{m}^{-3}$ to $10^{28}\ \mathrm{m}^{-3}$. This target is heated up to plasma conditions within very short time, such that the unconfined plasma cannot expand too quickly. Proposals for appropriate heating mechanisms are strong lasers, fast ion beams, or electric current (Z-pinch). Recent achievements are fuel gain, i.e. produced energy per incoming photon energy, above unity in laser fusion (heating to 60 eV within about 20 ns) [HCC$^+$14].

[2] A given amount of energy W of a charged particle can either be expressed as kinetic energy eU or as thermal energy $k_\mathrm{B}T$, where k_B is the Boltzmann constant, thus $1\ \mathrm{eV} \approx 11600\ \mathrm{K}$.

Magnetic Confinement Fusion

An alternative approach is to reduce the particle density and simultaneously increase the confinement time, i.e. to have a steady-state plasma. A typical thermonuclear plasma has $n = 10^{20}\ \mathrm{m}^{-3}$ to $n = 10^{22}\ \mathrm{m}^{-3}$; thus, the required confinement time is of the order of $0.1\ \mathrm{s}$ to $10\ \mathrm{s}$. While the detailed behaviour of plasmas is complicated, it is certain that it can best be confined in a magnetic field which is both

- toroidal to have closed magnetic field lines, since charged particles follow them, and
- helical with shear to avoid Grad-B drift on a large scale towards the upper and lower sides of the torus.

Despite there are other approaches (e.g. magnetic mirror devices such as the Gas-Dynamic Trap (GDT) in Novosibirsk), only this technology will be discussed in the following. See [RS83, Sta05, Sta10] for details. Mind that the plasma confinement is necessary for the reaction: Any significant leakage or otherwise undesired operating conditions will immediately lead to expansion, cooling, and recombination of the plasma and thus stop the reaction. This, as already mentioned, is different from fission plants, where chain reactions can take place if the main reaction is not conducted in a porperly controlled manner.

1.1.2 Plasma Confinement Concepts

The required magnetic field configurations can be achieved in two ways: either indirectly using a relatively simple coil configuration plus a magnetic field created by the plasma itself (tokamak principle), or directly using coils of complicated shape (stellarator principle), see Fig. 1.3. For the former, a plasma current has to be driven along the torus. In either case the coils for large fusion facilities have to be made of superconducting material to avoid ohmic losses, e.g. NbTi (ductile, but practical field limitation to 8-10 T) or Nb_3Sn (brittle, but larger superconduction region in B-T-j space).

Tokamak

The static component of the magnetic confinement in a tokamak is maintained by vertical, D-shaped toroidal field coils (TFC), with possible corrections by horizontal, O-shaped poloidal field (ring) coils (PFC). The dynamic component, the current drive, can be created by three methods (or a combination of them):

- Inductive heating: A strong central solenoid (CS), called ohmic heating (OH) transformer, is placed in the symmetry axis of the torus. While its magnetic field changes, an electric field in toroidal direction is generated, accelerating the charged particles. For current drive, the electric field must continuously point in the same direction, which means that the axial component of the CS field can either be increased or decreased during the time in which the plasma is confined. Thus, the inductive method only allows pulsed operation.

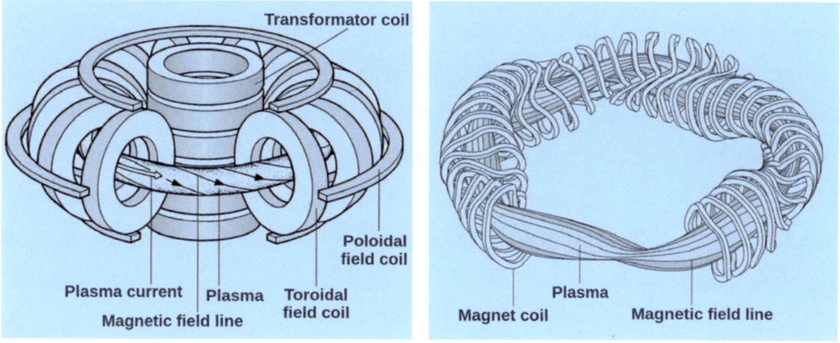

Figure 1.3: Magnetic confinement principles: Tokamak (left) and stellarator (right). See text for explanations. [IPP03], own translation.

- Collisional heating: The plasma is driven in mostly one direction using strong particle or electromagnetic beams, for example microwaves (electron cyclotron current drive, ECCD). This method requires that the particles/waves enter the plasma vessel through wall ports/antennas.
- Internal heating: Due to radial variation of the plasma pressure a spontaneous (non-inductive) plasma current can occur, termed *bootstrap current*. If this current is maximized inductive and collisional current drive can be reduced, even down to steady-state operation [RBB^{+}12].

Since each plasma is resistive to a certain extent, current drive will also always heat the plasma; and the collisional heating methods described below might additionally drive the plasma.

In tokamaks, due to the dynamic stabilization of the plasma, instabilities are likely to occur that might grow and eventually disrupt plasma confinement. The most prominent example are neoclassical tearing modes (NTMs), where the plasma temperature is lower than required. Thus, to mitigate these plasma modes, the plasma has to be heated locally.

The details of the confinement are complicated and depend also on plasma density, plasma pressure, and other factors. Notable current examples for tokamaks are ASDEX Upgrade (Germany), Tore Supra (France), JT-60 (Japan), DIII-D (USA), JET (Joint European Torus), and ITER, the latter being an unpreceded international collaboration between China, the EU, India, Japan, South Korea, Russia, and the USA. Apart from actual tokamaks there are some other tokamak-based concepts like Heliotron, Spherical Torus, or Reverse Field Pinch.

Stellarator

In contrast to tokamaks, stellarator-type devices do not require current drive, but their coil layout is more complicated and they generally need more space for the same plasma volume. Notable state-of-the-art examples are LHD (Large Helical Device, Japan) and Wendelstein 7-X (W7-X, Germany). Potential fusion power plants using the stellarator concept are referred to as *helical advanced stellarators* (HELIAS). See Fig. 1.4 for the main parameters of the world's most advanced tokamaks and stellarators in history.

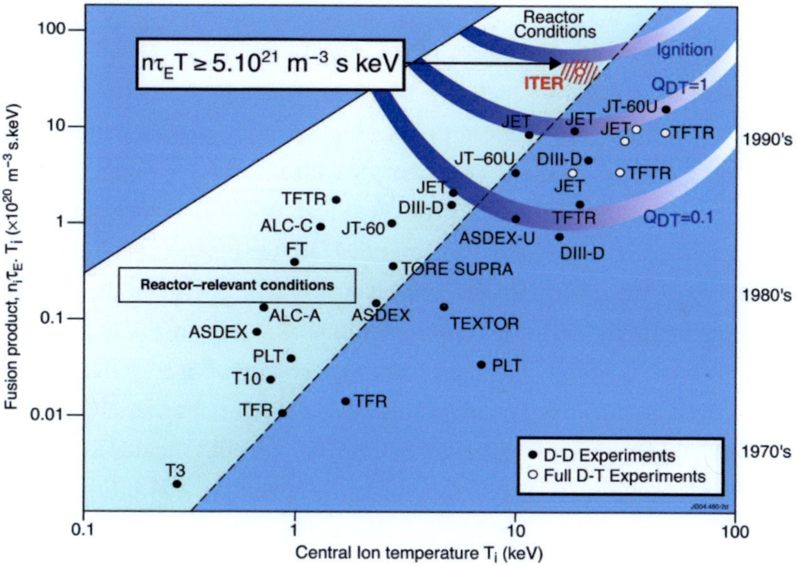

Figure 1.4: Achieved triple products and central ion temperatures of various plasma experiments [EUR11].

1.1.3 DEMO: The Way to a Fusion Power Plant

Apart from the magnets, experimental fusion facilities and power plants consist of the following main components [Sta10]:

- **Plasma vessel (vacuum chamber):** The low plasma density corresponds to a vacuum of about 10^{-5} Pa (10^{-10} bar) that can only be maintained if the plasma is placed in an evacuated and cryocooled vessel. Advantageous structural materials for the vessel and for most other parts are low-activation steel (Eurofer) or, perhaps in the future, silicon carbide (SiC).

- The innermost component of the vessel is called **first wall** and consists of individual panels, the **blankets**, which have to withstand (and absorb) heat and neutron fluxes. Furthermore, plasma particles can escape their magnetic confinement with a certain probability (especially during startup and shutdown) and hit the first wall, leading to erosion and to an influx of impurities into the plasma and thus reduce the plasma quality. In the blankets the kinetic energy of the outcoming neutrons has to be converted into heat and the required tritium to be bred; therefore, the blankets have to consist of at least five components: A shielding material (tungsten), a structural material (typically steel), a coolant (water and/or helium; maybe PbLi, liquid Li or LiBe), a breeding material (Li_2TiO_3 or Li_4SiO_4; maybe PbLi, liquid Li or LiBe), and neutron multipliers (Be or Pb). As one can see, one tries to combine breeding material (which has to be a lithium compound), coolant and/or multiplier. ITER will have test blankets in some of its upper ports in order to study different breeding blanket concepts, but will not breed tritium on a self-sufficient scale.

- **Divertor:** It is necessary to steadily clean the plasma from impurities and from helium ash. For this reason the outmost magnetic surface of a tokamak plasma has one (rarely more) X-points, where it intersects itself. Ions are then likely to escape the plasma at this position and hit nearby surfaces, from which they can be pumped outside the vessel. This system is called the divertor. The plasma-facing components of ITER divertors consist of carbon (planned to be replaced by tungsten).

- **Fuel cycle:** A significant part of the exhaust particles consists of deuterium and of the radioactive tritium which did not take part in the nuclear reaction and which are to be re-used in the plasma; especially the tritium at best without any loss. Thus, a dedicated system of in-situ analysis, purification, storage, and re-injection of both hydrogen species has to be designed and installed.

- **Diagnostics; remote handling:** Not only for experimental facilities like ITER, but also for demonstration and, later, for commercial reactors some degree of plasma diagnostics will be necessary. Quick and safe remote handling of components, e.g. exchange of blankets in *Hot Cells*, will also be of crucial importance.

The various **plasma heating** concepts will be explained in more detail in the next Sections.

So far, all existing plasma/fusion devices, including ITER, are experimental facilities, i.e. they are not designed to produce a significant amount of energy. This includes JET, which has however been operated with burning deuterium-tritium plasma. Fusion power plant concepts are at the moment studied in the EU, where they are referred to as DEMO concepts [FKW$^+$14] – short for *demonstration power plant*, since their purpose is to demonstrate that energy production from nuclear fusion is technically and economically feasible before commercial fusion power plants will be built. Parallel projects are underway in other ITER collaborators, e.g. China (FDS-II) [Wu08], Korea (K-Demo) [KIK$^+$15], the US (ARIES-AT) [NAB$^+$06] and Japan (SSTR) [KSN00]. DEMOs are currently planned for 2040-2050 [GS08, RBB$^+$12], usually as tokamaks, but power plants could also be designed as stellarators, e.g. in Germany if Wendelstein 7-X performs as expected [RBB+12]. The basic layout of a fusion power plant with emphasis on the fuel cycle is shown in Fig. 1.5. Typical figures for a 1.6 GW DEMO reactor are 50 MW bulk heating and current drive, 10 MW to 15 MW for magnetohydrodynamics (MHD) purposes (Neoclassical Tearing Mode (NTM) and sawtooth (ST) stabilization), and 50 % total ECRH efficiency, including transmission losses. Required EC frequencies are from 160 GHz upwards to 240 GHz, depending on the device's aspect ratio (torus radius : plasma radius), e.g. 150 GHz for initial heating and 180 GHz for current drive (bulk heating and NTM stabilization) for an aspect ratio of 2.6; 160 GHz/195 GHz for an aspect ratio of 3.1; and 190 GHz/240 GHz for an aspect ratio of 4. In Europe, there are two variants under discussion, a pulsed "DEMO1" and a steady-state "DEMO2" using more developed technology; both, however, intended as direct predecessors of commercial power plants [GBB+15].

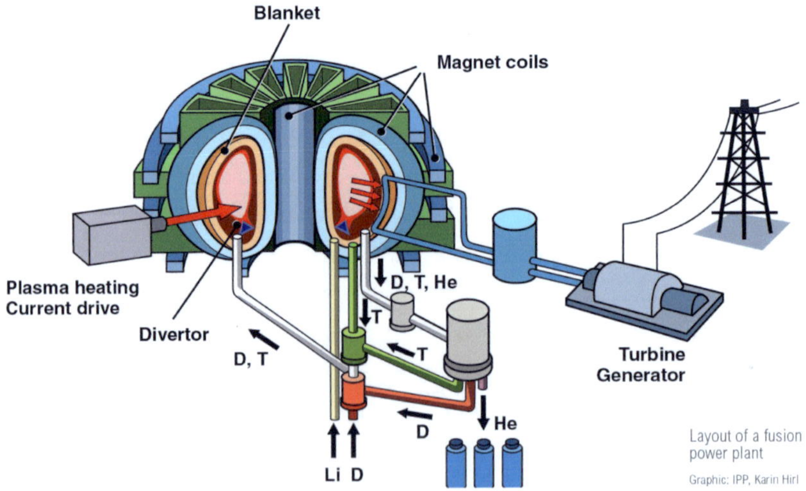

Figure 1.5: General layout of a future fusion power plant [IPP].

There are several issues which have to be solved porperly before DEMO can be built and operated. These include usage of stable low-activation materials, tritium self-sufficiency and safe breeding, stationary plasma conditions, and reliable, efficient plasma heating.

1.1.4 Fusion Plasma Heating

Before the plasma reaches the intended burning state it has to be heated externally. The so-called fusion gain factor Q describes the ratio of achieved fusion power to external heating power of a fusion plasma. Typical values for power plants are $Q = 20$ to 40 [GS08], where $Q = 1$ is commonly referred to as *break-even point* and, since the neutrons carry away $4/5$ of the energy, external and internal heating of the plasma are the same for $Q = 5$. There are four collisional plasma heating methods: neutral beam injection (NBI), electron cyclotron (EC) (resonance) heating, ion cyclotron (IC) heating and lower hybrid (LH) heating[3], competing with each other on the question which one would be the best for a future fusion power plant. Plasma heating with electromagnetic waves ("wave heating": IC, EC, LH) is based on resonant wave-particle interactions: there are resonance frequencies at which the plasma absorbs the waves efficiently, see Annex A.5. RF beams are polarized, can be generated far away from the plasma vessel and guided to it by waveguides and, if their wavelength is small enough, can be focused at certain spots in the plasma.

The four collisional heating methods in more detail are:

- **Neutral beam injection:** In order to cross the barrier imposed by the magnetic field, the net charge of injected particles (or compounds) has to be zero. Neutral high-energy hydrogen or deuterium beams are produced by accelerating ions (mostly H^-) and neutralizing them

[3] In the different experiments, the acronym of the heating method varies, e.g. Electron Cyclotron Heating (ECH), Electron Cyclotron Resonance Heating (ECRH), Electron Cyclotron Range of Frequencies Heating (ECRF), ... In this work, *ECRH* is chosen.

right before they enter the plasma vessel. Here they ionize and release their kinetic energy via Coulomb interaction; realistic values are 20 kA/320 keV for a 20 s deuterium beam [Sta05]. NBI is only weakly dependent on the plasma temperature. The disadvantage of NBI is that it needs a large space in the vessel wall and that the beams are so thick that they cannot be focused on plasma instabilities. The total efficiency from electric power of the NBI system to the heating power of the plasma is also an issue.

- **Ion cyclotron heating:** The ion-ion resonance is in the region of 30 MHz to 120 MHz. IC heating thus requires waves with vacuum wavelength of up to 10 m and, correspondingly, rather large antenna structures as well as contact of the antenna to the plasma to reduce the local wavelength. IC-relevant radiation comes from triodes or tetrodes. The Coulomb interaction with the plasma ensures that the electrons are also heated.

- **Electron cyclotron resonance heating:** The electron cyclotron resonance is of the order of 100 GHz to above 200 GHz, corresponding to wavelengths of around one to three millimeters. Kinetic energy is transferred to the ions due to Coulomb interaction. There are two plasma modes, ordinary (O-) modes (linearly polarized, with $\vec{E} \parallel \vec{B}$) and extraordinary (X-)modes. Gyrotrons are used to produce such microwaves with sufficient power and efficiency which are then guided to the plasma vessel by oversized waveguides. EC current drive is not very efficient; but since the wavelength of the radiation is small, the beams can be focused very well and can therefore be used for control of the plasma stability.

- **Lower hybrid heating:** There are two hybrid frequencies: lower hybrid ($\omega_{\mathrm{ion}} < \omega_{\mathrm{LH}} < \omega_{\mathrm{e}}$) with resonance frequency around 1 GHz to 8 GHz ($\lambda \sim 10$ cm), and upper hybrid ($\omega_{\mathrm{UH}} \gtrsim \omega_{\mathrm{e}}$). Application of

LH waves is not a very successful heating method, but the most efficient current drive scheme. Waves are produced by klystrons, transmitted by standard rectangular waveguides and launched with phased (i.e. internally structured) arrays.

In this work it is examined how an existing technology for plasma heating in ITER, namely gyotrons as microwave sources for ECRH, can be adapted to be suitable for DEMO. While the relevant regime of output power and frequency has already been touched in several publications, an in-depth physical design study of the main components for DEMO gyrotrons does not exist yet. This work is exclusively dedicated to coaxial-cavity gyrotrons, a technology that promises very high output power at high frequencies at the cost of an advanced manufacturing process. In parallel to this study, the possibility of using the hollow-cavity gyrotron concept is investigated at IHM [KAF+15a, TFK+15, KAF+16a, KAF+16c], which promises more robust gyrotrons with less output power.

1.2 High-Power Gyrotrons

1.2.1 Interaction Mechanism

Consider a physical system, consisting of at least one charged particle, in an excited state. This system might drop to a lower energy level by emitting a photon, either spontaneously or after interaction with another photon of the same energy, i.e. with a resonant photon. The second case is commonly termed *stimulated emission (of radiation)*. If several of such systems are placed nearby (i.e. form an ensemble, the gain medium), one incoming photon can trigger an avalanche of coherent photons; and if at least some of the photons are reflected back into the medium (i.e. if they are trapped in a resonator), one can ensure that most of the systems participate in that avalanche. By continuously replacing the systems which are in ground state or by re-exciting them (*pumping* the medium to obtain a population inversion), one can obtain permanent coherent radiation. This is the general maser/laser principle (Microwave/Light Amplification by Stimulated Emission of Radiation) [TBP13].

While in some of such devices the gain medium are atoms or molecules (e.g. gases or solids) and while the energy levels used are the respective orbitals, one can also use solely electrons in vacuum, gyrating in an external magnetic field, like in a cyclotron. These electrons have discrete energy states (Landau levels) due to their quantized angular momentum. In this case the pumping is done by a continuous input of higher-energy electrons through the cavity resonator. This is the principle of electron cyclotron masers (ECMs) [HG77]. ECMs have the obvious advantage that neither their frequency nor their energy or efficiency is constrained by material properties of some gain medium. Quantitatively the number of (almost equidistant) rotational states which can be occupied is much larger than in atomic/molecular lasers, and their energy difference is significantly lower, allowing photons with longer wavelengths – specifically, on the other side of the terahertz gap, i.e. with $\lambda \gtrsim 1\ \mathrm{mm}$.

In ECMs, the electrons radiate only if their phases are not equally distributed. But since the cyclotron frequency[4] Ω_C depends on the electron energy due to the relativistic factor γ [KBT04],

$$\Omega_\mathrm{C} = \frac{\mathrm{q_e} B}{\gamma \mathrm{m_e}} \tag{1.15}$$

($\mathrm{q_e}$ is the electron charge, B is the applied magnetic field and $\mathrm{m_e}$ is the electron's rest mass) and since the acceleration/deceleration of the electrons depends on their phase relative to the RF field $\vec{E}$ [Jac99],

$$P = \mathrm{q_e} \vec{v} \cdot \vec{E} \tag{1.16}$$

(P is the power transferred to the electron and $\vec{v}$ is the electron velocity), phase bunching occurs after some interaction time. During this process and with a properly tuned initial cyclotron frequency, the electrons will on average lose kinetic energy to the RF field via bremsstrahlung. The electrons can have either mildly relativistic or highly relativistic speed ($\geq$ 1 MV), the latter devices being specifically termed Cyclotron Autoresonance Maser (CARM). Basically, the electrons need to stay in phase with the RF field in their direction of travel, i.e. their velocity has to be similar to the group velocity (smaller than speed of light) of the field, fulfilling [Edg93]:

$$\omega(k_z) \approx s\Omega_\mathrm{C} + k_z v_\mathrm{gr} \tag{1.17}$$

Here, ω is the RF (angular) frequency[5], s is the cyclotron harmonic, k_z is the axial wave vector and v_gr is the group velocity in the electron guiding center's rest frame. This implies forward- or backward-directed

[4] In this work, *cyclotron* will be used synonymously for *gyro-* and *precession*, as in "cyclotron radius", "cyclotron period" or "precession frequency".

[5] As it is common in the gyrotron community, the term "frequency" will be used and corresponding values $f = \omega/(2\pi)$ will be given in this work, despite the fact that most formulas include angular frequencies ω. The only exception to this are window resonance frequencies, expressed as ordinary frequencies f.

fast waves, i.e. waves with a phase velocity larger than the speed of light (since $v_{\mathrm{ph}} v_{\mathrm{gr}} = c^2$), or, equivalently, with a refractive index $n := c/v_{\mathrm{ph}}$ smaller than 1. But there are also other suggestions, e.g. to operate a CARM in the Cherenkov regime using slow waves, which are usually associated with axial rather than azimuthal bunching. Mildly relativistic fast- and forward-wave ECM oscillators are called gyromonotrons or, shortly, gyrotrons [KBT04, Nus04, Edg93]. For typical gyrotrons the ratio between transversal velocity $v_\perp$ and axial velocity v_z of the electrons, the pitch factor g, is not much larger than unity. As a rule-of-thumb, $f(B) \approx 28\ \mathrm{GHz/T}$ for nonrelativistic electrons and $f(B) \approx 24\ \mathrm{GHz/T}$ for typical high-power gyrotrons ($U_{\mathrm{acc}} \approx 85\ \mathrm{kV}$).

Since the electron beam needs at least one dimension to pass through the cavity, there are two general cavity geometries for ECMs: plane-parallel (with a sheet beam) or axial (with a hollow beam), see e.g. the considerations in [DK81], of which the former are limited to sub-MW power [GGG+12]. For the axial configuration, there are two possibilities: either all electrons gyrate directly around the same axis (Large-Orbit Gyrotron, LOG [LB82]) or around axes which themselves are arranged around a central axis. If the electron orbits are far away enough from this purely geometric axis, one could also consider placing a metallic insert there, which would play the role of an additional cavity boundary.

Gyrotron cavities are open towards both ends, thus appear as parts of inhomogeneous waveguides. Electromagnetic waves within them will be at least partly reflected by any inhomogeneity, such as a narrowing downtaper in the direction upstream of the electron beam or a widening uptaper in the downstream direction, hence they can behave like being in a cavity resonator. The excited RF field in this resonator will appear as a number of eigenmodes, i.e. standing electromagnetic waves, which can in principle be calculated by either solving Maxwell's equations in differential form or using iterative methods, such as integration paths demonstrated in section 23-2 of [FLS62]. Usual high-power high-frequency gyrotrons are strongly overmoded, which implies that the spectral density, i.e. the typical distance between mode eigenfrequencies Δf relative to their frequency f, is high and that there can exist many modes above cutoff, leading to mode competition. By proper design only one mode is excited or it suppresses all other modes sufficiently and thus, the gyrotron can produce monochromatic output.

In the "small-orbit" design it makes sense to use a circular waveguide as the cavity, not only for manufacturing reasons, but also to ensure a homogeneous (i.e. everywhere minimized) ohmic loading from the RF field on the cavity wall. The interaction with the electron beam is highest if the electric field vector lies in the plane of the gyromotion of the electrons; thus, one is primarily looking for transverse electric (TE_{mpl}) modes in the cavity, where m denotes the azimuthal, p the radial and l the longitudinal index[6] (see Annex B.1). Since the cavity is axisymmetric those modes have a rotationally symmetric field maximum, and one would place the beam there (or, for $\mathrm{TE}_{m,1}$, close to the maximum); thus, the electron beam should at best be axisymmetric as well. Such beams also have the advantage of uniform charge distribution in azimuthal direction, reducing the risk of electron beam instabilities.

In hollow cavities with sufficiently small taper, the electric and magnetic fields on the axis must be unambiguous in magnitude and direction and can thus only be zero or point parallel to the axis for either TE or TM modes. In coaxial cavities [VZO76], i.e. cavities with a metallic rod along the symmetry axis, the second boundary condition on the inner rod removes this possible unambiguity. Therefore, it is possible to excite TEM modes in coaxial cavities; but they will be disregarded in this work due to their propagation (in axial direction) at the speed of light c. Nevertheless, the coaxial insert reduces TE mode competition (see [IKP96] and B.4 for details) and is therefore advantageous, despite the fact that it makes the gyrotron manufacturing clearly more complicated, as will be discussen in Chapter 3 in

[6] In this work, mode indices will be separated by comma if they are expressed as numbers or formulas, regardless of whether they are one-digit integers or not.

more detail. Figure 1.6 gives an example of a very high-order TE mode in a rotationally symmetric coaxial cavity (cross-sectional cut). For details of the coaxial insert in a gyrotron, see Fig. 1.11.

In a fusion power plant the power necessary for plasma heating, among other contributions, has to be subtracted from the generated power to obtain the net power of the facility. Considering ECRH only, gyrotrons are therefore required to operate at high efficiency in order to ensure high efficiency of the whole plant [RBB$^+$12]. Furthermore, the anticipated realistic output power per tube (between 1 and 2 MW) will be much smaller than the total ECRH/ECCD power needed; therefore, any decrease in total number of gyrotrons needed will relax the necessary effort for control, maintenance and exchange of individual tubes.

The general design of an advanced high-power, high-frequency gyrotron is shown in Fig. 1.7. A hollow electron beam is generated by a heated annular emitter in the temperature-limited regime. The electrons are accelerated towards the anode and enter the strong magnetic field of a superconducting magnet, where they follow the magnetic field lines. This increasing field compresses the hollow beam adiabatically (according to Busch's theorem, see Annex A.8), increasing the pitch factor of the electrons. The electrons now enter the cavity, where they lose a fraction – typically one third – of their kinetic energy, specifically of the energy corresponding to their angular velocity component. The electron beam now leaves the high-field region, where it decompresses. Finally, the electrons hit the collector. In the ideal case, the collector consists of one or more biased electrodes which decelerate the electrons before impact; thus, their kinetic energy is partially recovered and the total efficiency of the tube is increased in comparison with a non-biased collector. The microwave beam, on the other hand, leaves the cavity in the direction of the quasi-optical output coupler, where the axial component of its wavevector increases due to the slowly increasing launcher radius. The beam is then coupled out in radial direction and leaves the evacuated gyrotron via a mirror system and a suitable window.

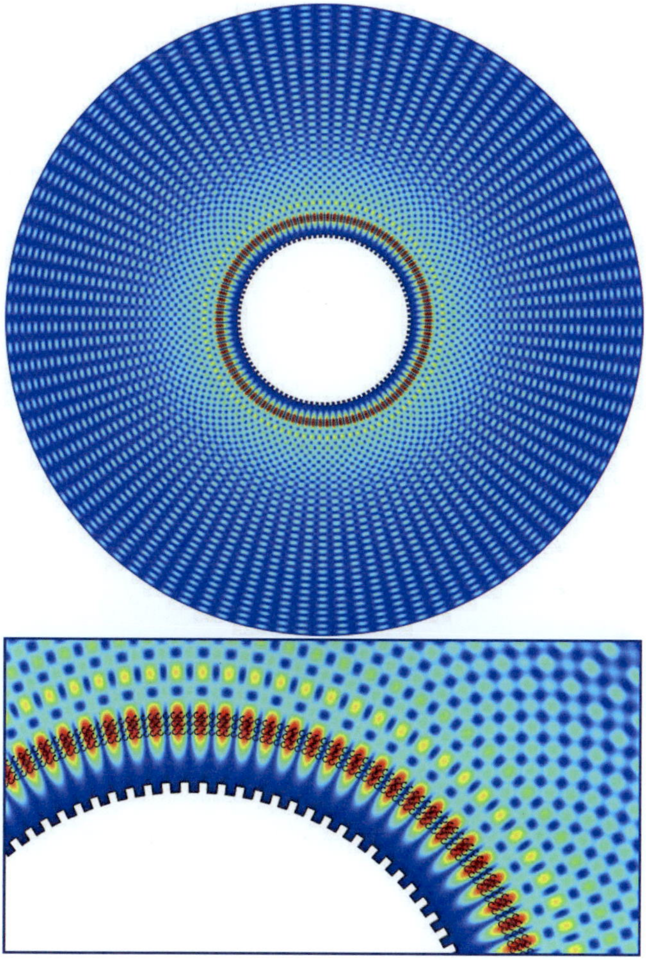

Figure 1.6: Operating mode $\mathrm{TE}_{49,29}$ in a waveguide with corrugated coaxial insert and with properly scaled electron beam (upper: full view; lower: detail). Color-coded is the amplitude of the electric field, where purple means small, green medium, and red large amplitudes. Mind that the actual azimuthal field maxima of the innermost ring are shifted by half a period versus the azimuthal maxima of the outermost rings, with a transition region around the fifth ring. As depicted, a guiding center spread of around three times the cyclotron radius needs to be considered in the design proposed in this work.

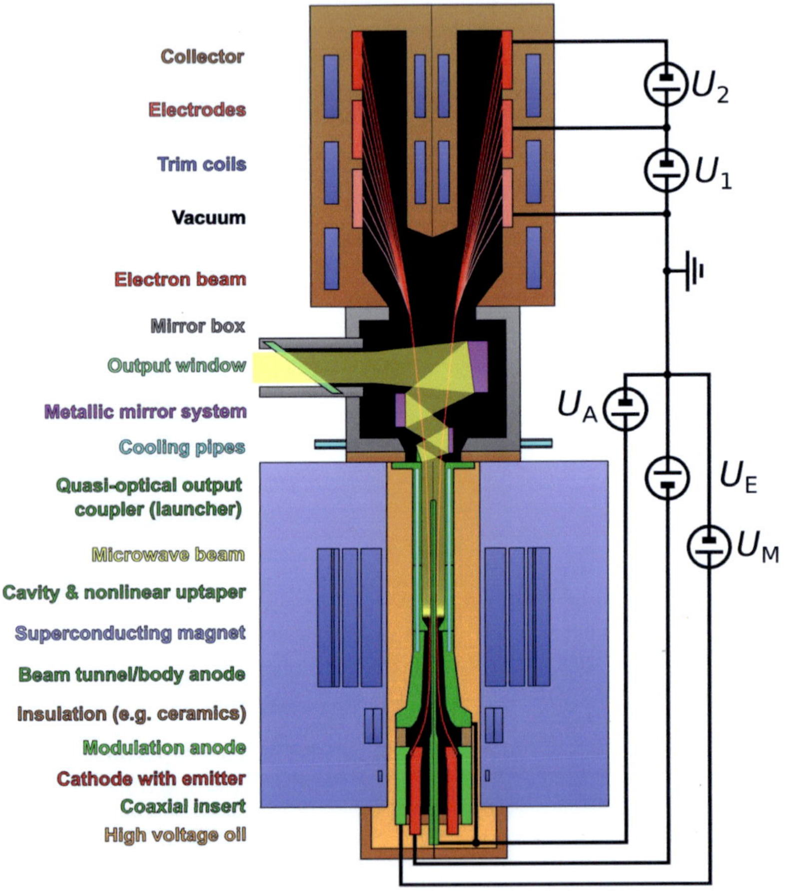

Figure 1.7: General design of an advanced high-power gyrotron. Height and diameter of the magnet are around 1 m. The electric connections to the electrodes usually require several dm space under the magnet, which is therefore placed on a rack. Red and green indicate negative and positive electric potentials, respectively: Insert, beam tunnel, cavity, and launcher are at the same potential. The Brewster-angle window is rotated by 90° around the RF beam axis to emphasize its geometry.

In order to find potential alternatives to gyrotrons for ECRH in fusion power plants one could consider different microwave tube designs (see [Bre91], section 3). In slow-wave devices such as klystrons, travelling-wave tubes (TWTs), backward-wave oscillators (BWOs), magnetrons, etc., the electrons pass the cavity structures very closely – of the order of the wavelength λ or even below $\lambda/10$ for TWTs – in order to ensure a high interaction efficiency. This implies that the electrons have to pass close to the metallic surfaces, which increases the necessity for cooling of the surfaces. Alternatively spoken, the dimensions of the delay line needed for slow-wave tubes become impractically small. For devices operated in continuous-wave (CW) mode, this limits the possible output power to values much lower than those of gyrotrons with the same frequency, see Fig. 1.8. Among fast-wave devices, gyrotrons, gyroklystrons, gyro-TWTs, peniotrons, and free-electron lasers (FELs, or ubitrons – *u*ndulated *b*eam *i*nteraction elec*tron* tube) exist.

- Gyro-klystrons differ from gyrotrons by having several interaction cavities in a row. Efficiencies can be comparable to those of gyrotrons, but frequencies and long-pulse power are lower [Thu16].

- In contrast to gyrotrons, gyro-TWTs lack a selective cavity and are therefore operated at low-order modes. As with gyro-klystrons, frequencies and output power are lower than for gyrotrons [Thu16]. However, recent conceptual studies exist to use a HF input in gyrotrons via the quasi-optical launcher (QOL) in order to improve mode-selection (*frequency* or *injection locking*), either from another device [BDN14], or as self-injection locking [MRR$^+$16]. Gyrotwystrons combine the properties of klystrons (input cavity) and TWTs (wideband output structure).

- Peniotrons use rectangular waveguide structures, which are disadvantageous due to the very high ohmic loading.

- FELs can reach extremely high power at very small wavelengths and have been considered for plasma heating [SJO+90], but efficiencies and pulse lengths comparable to those of gyrotrons have not yet been achieved (see [Thu16], sections 13, 14).

For a mathematical illustration of the resonance condition in various tubes, see A.9. Further information on vacuum tubes can be found in [EH89], and on lasers/masers in [KM69].

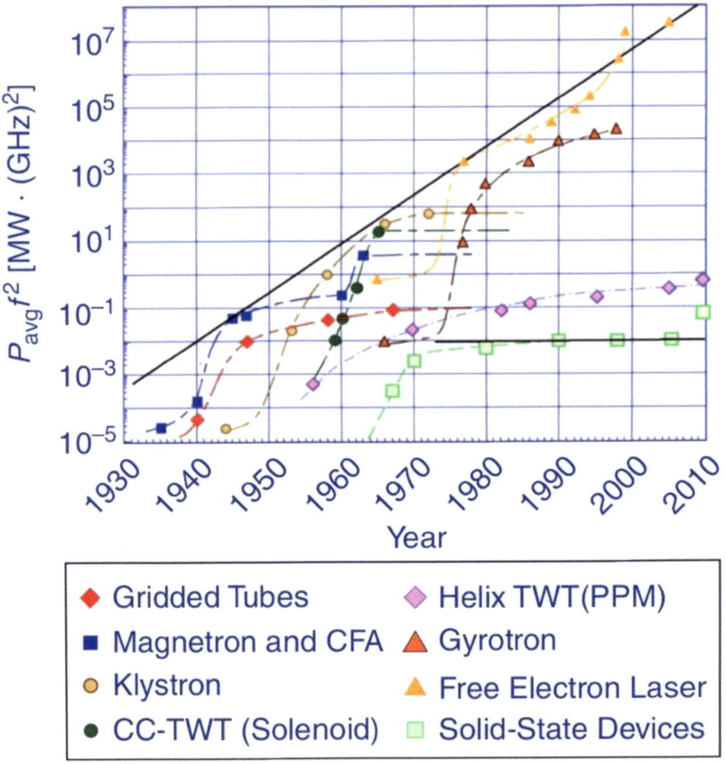

Figure 1.8: Historical development of Pf^2 of several amplifier types. For a given frequency, gyrotrons can deliver the second highest power after FELs, but at higher efficiency and pulse length than FELs. Reproduced from [QLP+09].

1.2.2 Key Components of High-Power Gyrotrons

One of the first articles presenting the basic gyrotron design as it is still considered today, although under the more general term *electronic maser*, was published by Soviet scientists in 1967 [GPY67] and had a focus on the bunching mechanism. The authors' goal was the same as it is still today: to obtain high-power millimeter radiation. In their review [FGPY77] ten years later the researchers from Nizhny Novgorod (then Gorky) could already refer to numerous articles on gyrotron theory and experiments. At that time, low-order modes such as $\mathrm{TE}_{0,1,1}$ and $\mathrm{TE}_{0,2,1}$ were employed to operate first- (i.e. fundamental, $s = 1$) and second-harmonic gyrotrons with wavelengths around 1 mm, efficiencies up to 50 % and output powers of about 20 kW. The challenges faced when increasing output power and efficiency of these vacuum devices were summed up by Gaponov *et al.* in 1981 [GFG+81], who also gave a good overview over the key components of gyrotrons. The possibility of coaxial-cavity gyrotrons and the use of gyrotrons for plasma heating is also discussed there. Zapevalov [Zap12] reviews the development of gyrotrons to state-of-the-art gyrotrons for fusion applications, which is the major driver for high-power gyrotrons. Thumm [Thu16] compiles yearly reports on high-power gyrotrons, and his compilation shows that the details of gyrotrons intended for the same purpose (e.g. heating in W7-X or ITER) vary considerably between the manufacturers Gycom (Russia), Toshiba (Japan), CPI (USA) and Thales (Europe). Nowadays, typical gyrotrons for fusion applications with SDCs deliver 0.8-0.9 MW at frequencies of 140-170 GHz with an efficiency of 35-57 % in CW operation ($t \geq 1000$ s). Gyrotrons with other frequencies, efficiencies or operating at higher harmonics are used for a broad range of applications such as materials processing [TLSJ15], detection of concealed radioactive substances [NPA+10], active denial systems [NRI09, SKK+13] or dynamic nuclear polarization (DNP) in nuclear magnetic resonance (NMR) spectroscopy [NBGT11].

Unless explicitly stated otherwise, this work only refers to MW-class gyrotrons with frequencies above 100 GHz in CW operation. Power, frequencies, or efficiencies higher than given in this work are mentioned in the literature, but only for the other values being significantly smaller; a list and discussion of those more exotic designs is given in Sect. 1.3.2.

Magnetron Injection Gun and Beam Tunnel

A magnetron injection gun (MIG) is the electron source of the gyrotron, owing its name from the similarity with the emitter of magnetrons, that is, producing a hollow beam parallel to a magnetic axis and showing significant space-charge effects[7]. Located in a magnetic field, an annular surface with low work function is heated from below and/or subjected to a strong accelerating electrostatic field ("cold" emission if no heating is applied), allowing the surface electrons to escape. These two operating conditions are termed temperature-limited regime (thermionic emission, described by Richardson's law) and space-charge limited regime, respectively. In order to obtain defined electron trajectories and thus to ensure high beam quality, gyrotrons are operated in the temperature-limited regime; and for simulations, the Richardson-Dushman-Schottky law [IL09] is commonly employed to describe the behavior of the MIG accurately:

$$j_{\mathrm{E}} = \frac{\mathrm{m_e e}}{2\pi^2\hbar^3}\left(k_{\mathrm{B}}T\right)^2 \exp\left(-\frac{W_{\mathrm{E}} - \sqrt{\frac{\mathrm{e}^3 E}{4\pi\varepsilon_0}}}{k_{\mathrm{B}}T}\right) \tag{1.18}$$

Here, j_{E} is the current density at the emitter, $\mathrm{m_e}$ is the electron mass, T is the emitter temperature W_{E} is the emitter material's work function and E is the applied electric field.

The emitter surfaces of cathodes intended for thermionic emission can be coated (for example, BaO on nickel or tungsten) or, for longer

[7] see IEC 60050, term 531-23-31

lifetime and reusablity, can be dispenser cathodes. B-type dispenser cathodes are usually made of porous tungsten, either impregnated with barium/calcium/aluminum oxide or attached to a barium reservoir. M-type dispenser cathodes are additionally sputter-coated with e.g. ruthenium, rhenium, osmium, or iridium. The porous tungsten can either be compressed powder, i.e. result from a stochastic process, or sintered wires, i.e. of controlled porosity. For practical reasons, the latter conflates with segmented emitters [MIP+12]. For both emitter types lifetime decreases considerably if emitter current density is increased moderately, see Fig. 1.9.

The initial properties of the annular electron beam are mostly influenced by the geometry of the emitter (radius, width, and slant angle) and by the direction and strength of the electric and magnetic fields at the emitter position. However, the shape of the magnetic field and of the electrodes nearby the emitter will also have an influence on the beam quality, especially on the pitch factor and on the statistical distribution of velocity components (velocity spreads).

If one uses just one cathode and one anode the electric field at the emitter can be modified only if the accelerating voltage is changed accordingly. A MIG with two anodes, one modulation anode opposite of the emitter and one main (or body) anode for the actual acceleration, allows to control beam energy and pitch factor separately.

Since a coaxial insert has to be fixed in the MIG region, coaxial-cavity and conventional-cavity gyrotrons require different MIG designs. However, during physical studies the mere existence of the insert does usually not have a significant influence on the key components of the MIG. From a technical point-of-view, in contrast, one has to consider cooling, fixation, and electric connection of the insert.

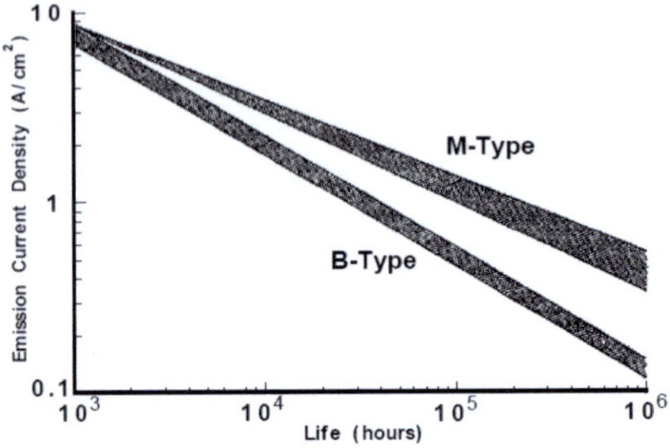

Figure 1.9: Emitter lifetime versus current density for B-type and M-type emitters. Reproduced from [Gil86], p. 132.

Several MIG designs have been proposed which can be classified according to the following criteria (other criteria are, of course, possible):

- diode or triode; in this work denoted "d-"/"t-" due to a lack of consistent nomenclature
- conventional or coaxial; "coaxial" denoted "C"
- conventional or with inverse emitter; "inverse" denoted "I"

This immediately leads to seven possible design types in addition to the conventional diode design d-MIG, see Fig. 1.10, most of which have been studied theoretically and/or experimentally:

- t-MIG: [BL86]
- d-CMIG: [Pio01]
- t-CMIG: [BIJ+09]
- d-CIMIG: [LMK+95] and [PBD+99]
- t-CIMIG: [RPG+16]

The only exceptions are conventional IMIGs, of diode or triode type, probably due to the fact that inverse MIGs work better in triode configuration and that the inner anode can readily be extended to a full coaxial insert. For a coaxial gyrotron in a so far unpreceded region of frequency and output power, e.g. of very high-order modes with eigenvalue $\chi > 120$, one would desire a triode MIG to study start-up scenarios and the effects of changes in beam parameters, especially the pitch factor, in more detail. An inverse MIG might, additionally to more efficient cooling, promise a more compact design due to the reduction of potential wells in its rear part [PR16] and thus a smaller magnet bore-hole diameter, but experience with inverse MIGs is rather limited at the moment. For this reason a t-CMIG design is most appropriate for the 2 MW 238 GHz gyrotron design, with the perspective to develop a suitable t-CIMIG in the future, e.g. based on [RPG$^+$16].

Several adverse phenomena can occur in the electron-optical system, especially the MIG and the beam tunnel. For example, since electrons can be reflected by magnetic fields decreasing along their path (magnetic mirror effect), electrons can get trapped between the magnetic mirrors around MIG and cavity if their pitch factor is too large. A second phenomenon are potential wells due to the combination of static electric and magnetic fields: if a magnetic field line crosses an electric equipotential line twice, electrons following the magnetic field line that have low kinetic energy (such as electrons from ionized residual gas) might not overcome the potential barriers corresponding to the equipotential line. Third, reflected or trapped electrons could hit the cathode surface at arbitrary points, releasing secondary electrons, which themselves could be reflected or trapped. As a consequence these three phenomena lead to uncontrolled accumulation of electrons in the MIG and tunnel region, which can lead to low-frequency oscillations in the gyrotron, to bombardement of anode surfaces [Zha16], or to arcing within the tube (Penning discharges) [PHG$^+$09, PHA$^+$10].

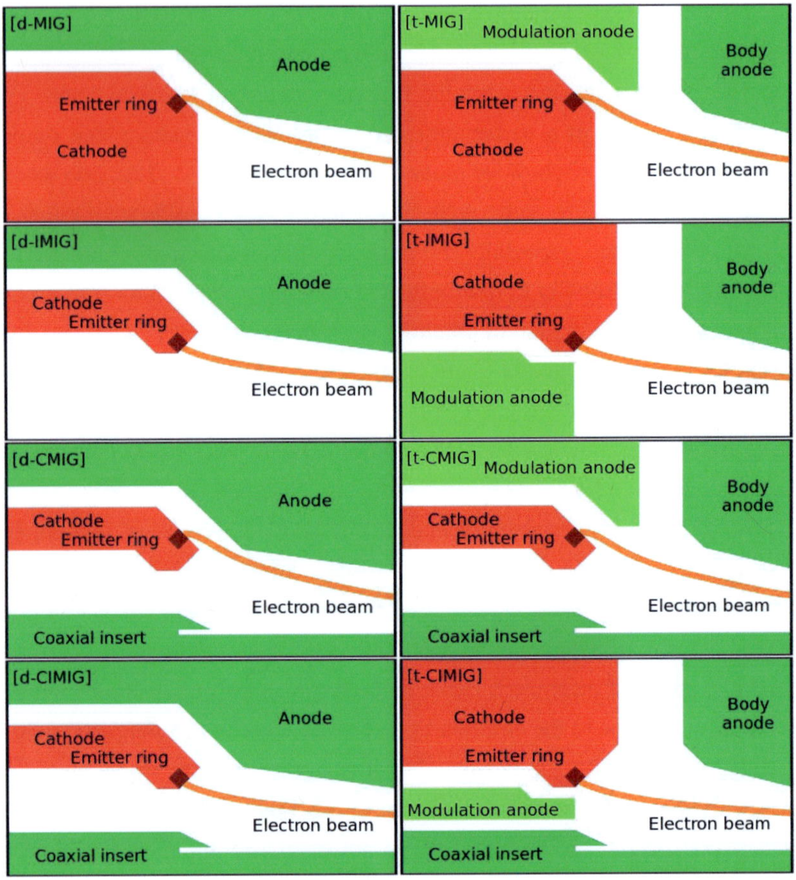

Figure 1.10: Schematic illustrations of the eight main MIG types as described in the text. The modulation anode potential is usually smaller than the main anode potential, but still positive with respect to ground (indicated by different shades of green).

Cavity

As it is well known there are two directions in which a non-axisymmetric mode can rotate with respect to the angular motion of the electrons: co-rotating and counterrotating, see Annexes B.1 and B.8. The coupling factor G_{mps} between hollow-cavity mode and electron beam is:

$$G_{mps}(r_\mathrm{b}) = \frac{\chi_{mp} \hat{B}_{z,m-s}(k_{mp} r_\mathrm{b})}{U_{\perp,mp}/(\omega_{mp}/k_{mp}^2)} \cdot J_{s-1}(k_{mp} r_\mathrm{C}) \tag{1.19}$$

where r_b is the electron beam radius, χ_{mp} is the mode eigenvalue, k_{mp} is the perpendicular wavevector, $\hat{B}_z$ is the field amplitude, $U_\perp$ is a normalization constant, J is the Bessel function, and r_C is the cyclotron radius; see also [Bor91]. The coupling itself is proportional to G_{mps}^2.

Since $\hat{B}_{z,m-s}(k_{mp} r_\mathrm{b}) = \hat{B} \cdot J'_{m-s}(k_{mp} r_\mathrm{b})$, the coupling to a chosen co-rotating mode ($m > 0$) can generally be stronger than to the corresponding counterrotating mode ($m < 0$) due to the larger global maximum of $J'_{|m-s|}$ compared to $J'_{|m+s|}$, provided $s > 0$ by definition. Physically, this inequality can be explained as follows: Each electron orbit can be divided into two zones: the zone where the electron is at a position more distant from the cavity axis than its guiding center (abaxial zone) and where it is closer to the cavity axis (adaxial zone). Clearly, the abaxial zone always covers a larger angle ($> \pi$) than the adaxial zone ($< \pi$). For one beamlet, i.e. an electron bundle with the same guiding center, more electrons are thus in the abaxial zone and, equivalently, stay there for a longer time. For co-rotating modes the field maximum is located at the abaxial position, while for counterrotating modes it is at the adaxial position, which accounts for the different coupling.

The three means of basic selection which mode shall be excited in the gyrotron cavity are electron beam radius, cavity radius, and cyclotron frequency. Thus, for one chosen mode and a fixed electron beam radius in a given cavity, all modes which have similar eigenvalues and similarly large or larger coupling factors are the main competitors of that chosen mode because they can be excited as easily. From Eq. 1.19 one can readily deduce that for co-rotating operating modes TE_{mp} (at least for typical gyrotron modes with $h_{mp} \geq 0$, see Annex B.1), the main competitors are the counter-rotating modes with smaller $|m|$ and larger p: $\mathrm{TE}_{(-m-3),(p+1)}$ and $\mathrm{TE}_{(-m-2),(p+1)}$ [Ker96]. Other modes that have a strong coupling to the beam and are therefore serious competitors are the azimuthal and radial neighbors of the main mode as well as the mode $\mathrm{TE}_{(-m-4),(p+1)}$, giving in total the ten modes listed in Table 1.1 to be at least taken into account in multi-mode simulations (such as for $\mathrm{TE}^{\mathrm{co}}_{28,16}$ in [Ker96], section 2.3.5). Note that for a caustic radius [Vaj69] of $C_{\mathrm{K}} \equiv m/\chi \approx 0.3$ (equivalently, $p/m \approx 0.6$), which corresponds to typical coaxial-cavity designs, the relative eigenvalues of the main competitors are symmetric to the main mode (0.996 and 1.004, respectively), i.e. the main mode is farthest separated from equally strong competitors. The set of coupling factors over eigenvalues is termed "coupling spectrum". Examples for coupling spectra are given in Fig. 2.2. One can well distinguish between modes with a coupling above approximately 60 % of a chosen mode which are alternative candidates for operating modes, but also candidates for competing modes, and modes with a coupling below 60 % which are much less likely to take part in the gyrotron interaction.

Optimum operation of a gyrotron requires not only the desired mode and the beam at proper position, but also optimized beam parameters, especially beam voltage and pitch factor. A certain starting current can be assigned to each mode and a given beam voltage U_{b}, being the minimum current at which the mode can be excited (*soft-excitation* or *supercritical Hopf bifurcation regime*). However, the parameter regions where modes can exist generally overlap, which means that for exactly the same cavity geometry and beam parameters, several modes can exist. If one of those modes has already gained some energy, it will usually suppress the excitation of the others, at least to a large degree, as long as it can remain stable.

Table 1.1: Co-rotating mode TE_{mp} and its main competitors with their relative eigenvalues (approximation for high m, p and $p/m \approx 0.6$) and relative caustic radii ("larger" and "smaller" being very close to the main mode C_K, while "small" differs more). The fourth column lists possible methods to suppress unwanted modes, see text for further discussion.

Mode	Relative χ	Relative C_K	Possible suppression
$\mathrm{TE}^\mathrm{co}_{m,(p-2)}$	0.96	larger	
$\mathrm{TE}^\mathrm{co}_{m,(p-1)}$	0.98	larger	
$\mathrm{TE}^\mathrm{ct}_{(m-4),(p+1)}$	0.988	small	
$\mathrm{TE}^\mathrm{co}_{(m-1),p}$	0.992	smaller	rippled outer wall
$\mathrm{TE}^\mathrm{ct}_{(m-3),(p+1)}$	0.996	small	coaxial insert
$\mathrm{TE}^\mathrm{co}_{m,p}$	1.000	—	—
$\mathrm{TE}^\mathrm{ct}_{(m-2),(p+1)}$	1.004	small	coaxial insert
$\mathrm{TE}^\mathrm{co}_{(m+1),p}$	1.008	larger	rippled outer wall
$\mathrm{TE}^\mathrm{co}_{m,(p+1)}$	1.02	smaller	
$\mathrm{TE}^\mathrm{co}_{m,(p+2)}$	1.04	smaller	

This effect can be used to leave the soft-excitation regime to a nearby region (with different U_b) in which the mode could not be excited directly (*hard-excitation* or *subcritical Hopf bifurcation regime*) [KBT04]. Consequently, the operating mode of a gyrotron is in general not a state function of the input parameters, but also depends on the time-evolution of the beam parameters – the gyrotron displays a hysteresis [DII$^+$03]. Since at least the beam current must change continuously from 0 A to the design value, one cannot jump immediately to the design parameters of the gyrotron, but must follow a path in the parameter space which in the end excites the desired mode (and perhaps other modes in advance). In many cases the maximum interaction efficiency can even be reached in the hard-excitation regime only, so the described mechanism has to be used. In this context, excitation of the correct mode and maintaining operation at high efficiency can be difficult in practice because the beam parameters, especially voltage, can be controlled only with finite accuracy.

The less dense the mode spectrum is, the more likely it is that the proper mode can in the end be excited. It is therefore very useful to find means that suppress the onset of unwanted modes, i.e. increase their starting current. A properly designed coaxial insert, see Fig. 1.11, can suppress the counterrotating competing modes, see [Ker96] and Annex B.4.

As already mentioned, the typical gyrotron cavity consists of a downtaper section (narrowing in downstream direction), a straight midsection and an uptaper section (widening in upstream direction). Usually, parabolic smoothings between the three sections reduce mode conversion, as does a nonlinear uptaper which provides a smooth transition between the cavity and the (larger) launcher entry [Ker96]. For cavities with corrugated insert the general assumption is that modes are either almost unaffected by the inner rod and therefore behave like hollow-cavity modes (practically no change in eigenvalue or caustic radius), or that they do not play a role during the interaction because they are suppressed by the insert. Therefore, most basic considerations apply equally to hollow-cavity and to coaxial-cavity gyrotrons. For a certain coaxial-cavity design, this assumption has of course to be verified.

The uptaper extends outside of the interaction region and acts here as an overmoded cylindrical waveguide to direct the RF beam away from the cavity. Nonlinear widening minimizes reflections back into the cavity and mode conversion, while the axial wave vector component increases (see Annex A.9). The downstream end of the nonlinear uptaper is typically 7 % larger than the cavity radius.

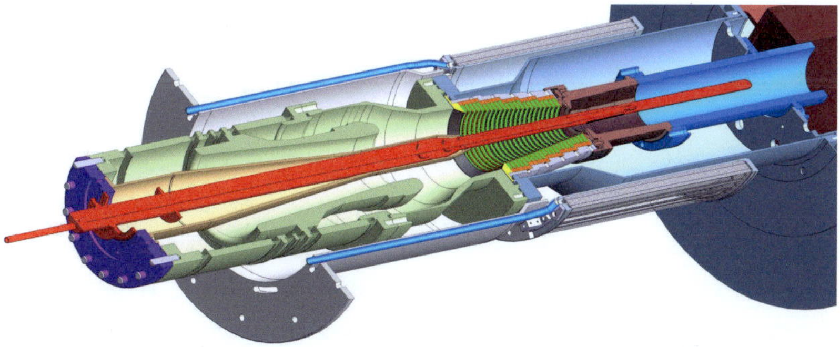

Figure 1.11: Lateral section of a coaxial-cavity gyrotron [PDD+05]. The coaxial insert (red) is supported (lilac) in the MIG and extends through beam tunnel (green/orange), cavity and nonlinear uptaper (brown) into the launcher (blue). According to present understanding, the actual end position of the insert within the launcher region is of minor relevance for gyrotron performance.

As with every metallic surface hit by microwaves, the cavity surface (and, if present, the insert surface) heats up and therefore has to be cooled continuously during steady-state operation. The ohmic losses of a TE mode in a cylindrical cavity can be expressed as:

$$P_\Omega = \sqrt{\frac{2\,\varepsilon_0\,\omega^3}{\sigma}} \cdot \frac{W/\chi}{1-(m/\chi)^2} \tag{1.20}$$

Here, χ and m are mode eigenvalue and azimuthal index, respectively, ω is the RF angular frequency, σ is the cavity material's conductivity and W is the stored energy in the cavity. For a derivation, see Annex B.7.

Typical means of cooling for the cavity as well as for other gyrotron components are Raschig rings or hyper-vapotron cooling. Possible effects of inhomogeneous cooling are described in Section 3.5.

Superconducting Magnet

The strong magnetic field in the cavity has to be provided by a surrounding superconducting magnet (SCM). In the interaction region, i.e. the mid-section of the cavity, the field should be maximum, and decrease sufficiently before and after this region in order to avoid possible Before- and After-Cavity Interactions (BCI [Sch15]/ACI [Cho14]), i.e. unwanted waves which are excited if their resonance conditions are unintentionally fulfilled. Since the electrons largely follow the magnetic field lines, the actual shape of the field from emitter to collector is of importance for a detailed design. If the magnet consists only of one coil or of an assembly of concentric coils with the same current and winding chirality (a *main coil*, MC), the magnetic compression, i.e. the ratio between the magnetic flux densities at emitter and cavity, depends on the distance between these two. However, if coils with opposite winding are added around the emitter region (*compensation coils*, CC, or *bucking coils*), the local field and likewise the required gyrotron length between emitter and cavity can be greatly reduced. If additional coil systems are fed by independent circuits one obtains the flexibility to control the shape and strength of the field at the emitter (*gun coils*, GC). Of course, additional coil systems serve both purposes to some extent, hence a distinction between CC and GC is to some degree ambiguous. In practice, magnets for the same class of gyrotrons can have various designs, see e.g. the magnets for the Japanese [SKK$^+$09, HKO$^+$08], Russian [SFK$^+$06], and European ITER [CAB$^+$13] and W7-X [KDK$^+$00] gyrotrons.

In Fig. 1.12 the lateral section of a gyrotron inside its magnet as well as the schematics of a European-style magnet coil configuration with corresponding current directions (assuming the same helicity for all coils) are shown.

On a large scale, two materials for low-temperature superconducting (LTS) wires are available: NbTi and Nb_3Sn. As already mentioned in the context of fusion plasma confinement, NbTi is ductile but can withstand local fields of up to only 15 T, while Nb_3Sn is brittle but can withstand fields up to 30 T [GCD+06]. Critical parts of the coil structure, where the local magnetic field is largest, are the edges of the windings. Thus, the achievable central field of SCMs with large bore-hole and NbTi-only coils depends on the winding, but it is typically below 10 T. In SCMs for powerful high-frequency gyrotrons, Nb_3Sn is therefore advantageous, especially for inner windings. Modern superconducting magnets mostly use cryogen-free cooling [HMHH05, Goo14].

If one could assemble a gyrotron in two subsystems (e.g. MIG – beam tunnel – cavity – launcher, and mirror box – window – collector) and insert them from both ends into the magnet before final assembly, the required magnet bore-hole diameter would be determined by the narrowest part of the gyrotron, which is the cavity region (see Fig. 1.11 or 1.12). Unfortunately, after final assembly, a gyrotron has to be baked out in order to achieve the necessary ultra-high vacuum of less than 10^{-8} Pa [KSM+04]; thus, the superconducting magnet would be deformed or its coils damaged when subjected to high temperatures. Therefore, the magnet has to supply enough space for the gyrotron to be inserted from either direction after it has been baked out. For this reason, the minimum bore-hole diameter is determined by the maximum diameter of the MIG. On the other hand, due to the very nature of the windings, the required cryostatic environment in the magnet and the strong magnetic forces, there is no known method to build magnets with flexible diameter (e.g. a "resting position" with wide bore-hole and an "operating position" with narrow bore-hole; or magnets that can be longitudinally split and recombined).

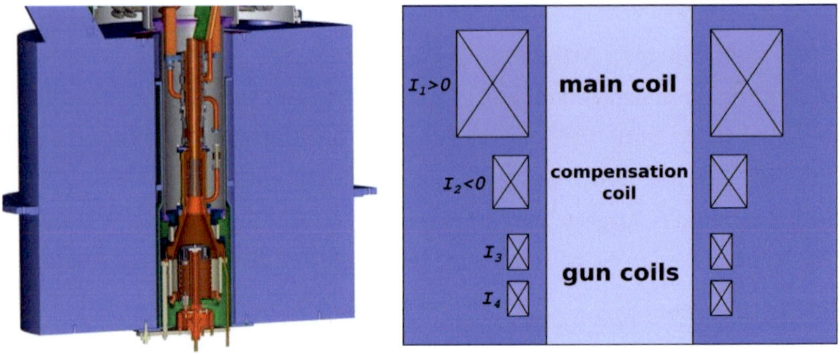

Figure 1.12: Gyrotron magnet with a gyrotron inserted from above (left); and EU-style coil schematics (right).

Quasi-Optical Output System; Launcher

Since the direction of the electron beam cannot be changed without using a very strong magnetic field, there are two methods to separate the RF beam from the electrons after both leave the nonlinear uptaper. First, the upper waveguide can be used as a collector for the beam electrons, leading to an axial RF output. This is the original method, and it has lately been used in the high-efficiency gyrotron described in [MIP$^+$12]. The second method typically employed for fusion gyrotrons is to couple out the RF beam in radial direction [VO74]. In order to achieve this, the inner surface of the output coupler has to be corrugated or deformed such that it acts as a line of focusing quasi-optical mirrors – due to the nonvanishing k_z (or, Poynting vector $\vec{S}$), the RF wave can now be understood as a beam climbing up the output coupler and being focussed successively. The launcher of a gyrotron designed for a co-rotating operating mode would be designed specifically for this rotational direction and not be appropriate for the counterrotating mode.

When the transformation is complete, the RF beam is launched radially. Since a TE mode was excited in the cavity, the launched beam is horizontally polarized, provided the gyrotron is upright. Final focussing, phase-correction and transformation into a Gaussian mode $\mathrm{TEM}_{0,0}$ (therefore, another name for the QOL is *Quasi-Optical Mode Converter*) is done by two or three separate mirrors, either all located inside the gyrotron's mirror box, see Fig. 1.13, or some outside it [SKT+97]. There are different sub-types of launchers, namely mirror-line launcher [VO74], helically deformed launcher [DKM+92] and hybrid launcher [JGTJ15].

Focussing of the beam strongly depends on the caustic radius of the mode to be transformed, since the caustic radius determines the inclination angle of the transformed beam, which itself determines the boundaries of the mirror line. Therefore, if one wants to minimize stray radiation in a gyrotron, the launcher of a multi-frequency gyrotron restricts all possible modes to those which have (nearly) the same caustic. Initial studies of a launcher designed for cavity mode $\mathrm{TE}_{49,29}$ and used for transformation of modes $\mathrm{TE}_{42,25}$ and $\mathrm{TE}_{35,21}$, as well as for $\mathrm{TE}_{43,15}$ with $\mathrm{TE}_{37,13}$/$\mathrm{TE}_{31,11}$, showed that constant caustic radius is quite sufficient for multi-frequency operation, even if the wall deformations/corrugations are optimized only for the highest mode in both cases [Jin13, JK15].

The microwave beam now has to leave the vacuum of the gyrotron to enter an evacuated transmission line, a line of mirrors in air, or another system to be guided to the plasma vessel. The most simplest way, however still challenging, to ensure vacuum inside the gyrotron is a single-disc vacuum window. Apart from being leak-tight and mechanically stable, such windows have to withstand up to 2 MW transmitted RF power, which is possible only with appropriate cooling. If the window thickness corresponds to multiples of $\lambda/2$ of the incident radiation, reflection into the gyrotron is minimized for the corresponding frequencies (but absorption is maximized) [BDKT97]. At present, windows made of chemical vapor-deposited diamond (CVD-diamond) have the best mechanical, thermal, and optical properties and are widely used for gyrotrons. Typical cooling allows a bandwidth of approximately 1 GHz around the window resonance frequency.

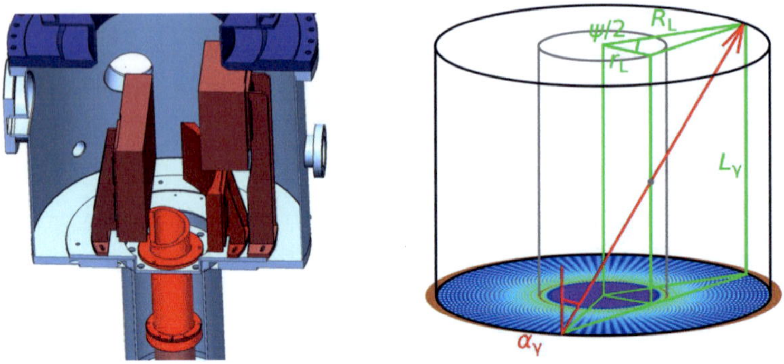

Figure 1.13: Typical arrangement of the quasi-optical mode converter (launcher; red) and mirrors (dark red/brown) of a gyrotron (left). Definitions of Brillouin angle α_γ, Brillouin length L_γ, spread angle ψ, launcher radius R_L and launcher caustic radius r_L (right). [DKM+92, Fla12, Jin13]

In order to increase this bandwidth, more complicatied window systems such as Brewster-angle windows [TAB+01] or tunable windows are under study [LDM+12]. Brewster-angle windows drastically reduce reflections back into the gyrotron over a broader range of frequencies because of the linearly polarized microwave beam coming from the mode converter. The window on the other end of the beam duct, in the RF launcher towards the plasma vessel, must have the same (or better) optical properties as the gyrotron window. [KBT04]

The radial output is not limited to one beam; indeed, there are several gyrotron designs in which the original beam is split into two beams before leaving the launcher in radial direction, see e.g. [PBD+98, Ber11] or the "Two-eyes gyrotron" [GDZ+96]. This of course requires twice the number of mirrors and two output windows, alongside means to ensure proper separation of the beams. As long as present window technology does not permit microwave windows that can withstand significantly more than 2 MW CW, all gyrotrons with higher power as well as 2-(or maybe even 1-)MW gyrotrons with less window cooling capacity have to be equipped with a dual-beam output system. On the other hand, this means that any realistic single-window gyrotron design can be restricted to 2 MW.

At all stages of RF generation and focussing, stray radiation is created alongside the main microwave beam, usually totalling a few percent of the generated power. The inner surfaces of the gyrotron barely absorb radiation, thus it accumulates basically uniformly inside the tube – mainly in the mirror box –, but it can also be reflected back into the cavity and disturb interaction [PDD+05].

Collector

After interaction the spent electrons leave the region of high magnetic field and reach the collector. Their residual kinetic energy, i.e. the energy not transferred to the RF field, makes up typically 65 % to 70 % of the energy gained by the acceleration. Upon impact on the collector surface the electrons would collide with the solid material and their kinetic energy would be transformed into heat. This resistive behavior of the vacuum-metal surface can be reduced if the electrons are decelerated by a depression voltage before impact, at best until zero velocity.

However, since the electrons have a rather broad energy distribution and since reflection of the electrons must be avoided for stable gyrotron operation, the depression voltage of a single electrode must correspond to the lowest-energy electrons. This means that most of the electrons still carry a significant amount of kinetic energy before impact, typically 50 % to 55 % on average. Therefore, fusion-relevant gyrotrons with single-stage depressed collectors (SDC) rarely reach efficiencies above 50 % [Thu16].

The most promising solution to increase collector efficiency is to sort the electrons by their energy, using static electric and magnetic fields, and collect them at various electrodes with proper depression voltages. Possible sorting mechanisms are $\vec{E} \times \vec{B}$ drift [PHA+08, PWI+16] and nonadiabatic transitions of the axial magnetic field [WPI+16a, WPI+16b], both of which use the fact that electrons do not gyrate strictly on helical trajectories around magnetic field lines. Such collectors could have efficiencies substantially above 70 % and could be used to increase the total gyrotron efficiency to 60 % or even more. Collector design concepts can be combined with basically any MIG and cavity concepts; therefore, a collector design is not part of this work. A power supply compatible to highly efficient Multi-Stage Depressed Collectors (MSDCs) is currently under development at KIT [SFK$^+$15].

1.2.3 Frequency Tunability and Multi-Frequency Operation

Future fusion power plants are foreseen to operate at a fixed magnetic field with optimized plasma temperature and density profile. Hence, the correlation between radial position within the plasma and local resonance frequency is known. One can then fix the angles and positions of the launchers in such a way that, for the mitigation of NTMs, the required position in the plasma can be aimed at by just changing the microwave frequency. A few frequency steps of a few GHz within roughly one second each are sufficient to replace steerable RF launchers by fixed ones in DEMO. This can significantly reduce the design effort and vulnerability of the RF launchers in fusion power plants [ZT05, GGR$^+$15].

There are several possibilities for frequency tuning. The most intuitive is probably to change the cavity radius R_O, since the (cutoff) frequency f of the mode depends via its eigenvalue χ on the cavity radius: $f = c\chi/(2\pi R_\mathrm{O})$. Indeed, there is a frequency shift due to thermal expansion of the cavity during long-pulse operation by a few hundred MHz [Sch15]. Unfortunately, any controlled mechanical change of the cavity dimensions, e.g. using a split cavity [BCD$^+$84], by significantly more than this amount is not possible in high-power gyrotrons, since a RF field on the cavity walls will damage any small gaps, edges etc., and since the deformable cavity still has to be cooled. Cavities with piezoelectric shape control [SI08] cannot cover a sufficiently large bandwidth either. During ramp-up the process of space-charge neutralization leads to an additional decrease in frequency of the same order.

In coaxial gyrotrons, the mode eigenvalue also depends on the radius of the coaxial insert. Since the insert has to be tapered anyway (see Annex B.4), it could be mounted such that it can be moved in axial direction, leading to a varying insert radius at the cavity reference point [IDT08, DM98], maybe even combined with a tapered cavity [GLMK08]. Yet, the possible change in eigenvalue and thus in frequency is small and the alignment of coaxial inserts in gyrotrons is a challenge even for metallic rods intended to be fix [KDK+01].

Small-scale tunability for one particular mode is also possible by slight variation of the accelerating voltage and/or the magnetic field in the cavity B_z, since both lead to a change in the electron cyclotron frequency and thus in the resonance condition (see A.9), an effect which is called frequency pulling [KDW+84]. Larger variations of accelerating voltage and/or magnetic field will lead to a jump to another mode, termed frequency step-tuning. For this, variation of B_z has been considered in the beginning (see e.g. [Edg93], chapter 11; and [DP90]) which has a range of up to 35 % [KBT04], but fast variation of the magnetic field is difficult especially for high fields. Therefore, the best tuning can be achieved by variation of the accelerating voltage U_{EA}, perhaps also by varying U_{EM} in the case of a triode MIG. For optimum output power and efficiency, however, both B_z and U_{b} should be varied. Typical variation ranges are then ± 5 %, see e.g. [KBT04], table 2.1. The operating mode will then usually be changed by only one of its indices – axial, azimuthal or radial – where the azimuthal index is most common and preferred. For gyrotrons with cavity radii around 2 cm to 3 cm, the typical frequency step size has a fixed value between 2 GHz and 3 GHz, respectively (see Annex B.9), or higher in the case of a change in the radial index.

During frequency step-tuning in the hard-excitation regime it is much easier to go downwards in frequency than upwards [KAF+16b]. A single gyrotron would have to be almost switched off and on again to obtain a higher frequency, which might be time-critical. But since in a fusion reactor several gyrotrons would be used per in-vessel launcher, those gyrotrons could either be operated at individually assigned frequencies or be controlled in a revolving fashion: the frequency needed f_P will be supplied by the gyrotron with the smallest frequency larger than or equal to f_P, thereby minimizing reaction time, while one of the gyrotrons currently at a small frequency f_min is switched off and prepared to operate at the maximum frequency f_max.

Apart from these methods, one can modify the axial field profile for fine-tuning by increasing the axial index of the mode [HBG+05] or using a tapered cavity [BNP76]. In any case, as soon as frequency shifts of more than around 1 GHz appear, simple single-disk windows are no longer appropriate since the deposited heat at the unmatched frequency can no longer be dissipated in CW operation. In order to broaden the transmission bandwidth, more complex systems such as the abovementioned Brewster-angle windows [BDKT97], double-disk windows [WGL+08], or travelling-wave windows [LDM+12, NTP14] are considered.

Additionally to this continuous or step-frequency tunability, multi-frequency operability, i.e. the possibility to operate the same gyrotron at significantly different frequencies, one at each time, can make the gyrotron useful for different purposes in the same facility (start-up, bulk heating, current drive, plasma stabilization), or even in several facilities. (Therefore, the term “multi-purpose” is also used for this property.) If a single-disk window is used, the operating frequencies f_n have to be integer multiples of the window resonance frequency f_1, where the reflection of the wave back into the gyrotron is minimal. Provided a loss tangent of approximately zero and $\mu_\text{rel} = 1$, these resonances are related to window thickness D_W and relative permittivity ε'_rel via

$$(2f_1 D_W)^2 \varepsilon'_{rel} = c^2 \tag{1.21}$$

$$f_n = n \cdot f_1 \tag{1.22}$$

For CVD-diamond windows, ε'_{rel} is given as 5.67 ± 0.01 [KBT04]. The lower limit of the windows' thickness is given by their mechanical stability, and the upper limit by their thermal conductivity and their price; typical windows then have a thickness between 2.25 mm [ZBD+04] and 1.50 mm [ABG+15], which results in resonance frequencies between 28 GHz and 42 GHz. Considering this, Fig. 1.14 shows possible multi-frequency options: The horizontal axis shows typical lowest frequencies of multi-frequency gyrotrons, and the vertical axis shows possible higher frequencies. Lines show possible higher frequencies to a lowest frequency; and same colors correspond to same window thicknesses. One can see that the restrictions are significant; for example, a gyrotron with basic frequency 130 GHz can only have higher frequencies at 162.5 GHz, 195 GHz, 227.5 GHz etc. During the design phase of a multi-frequency gyrotron, variation of the lowest frequency while remaining at the same relative window resonance n (i.e. remaining at the same color in Fig. 1.14) can basically be achieved by scaling the existing gyrotron design. On the other hand, changing the main frequency to a value that cannot be provided by a window at the same resonance (i.e. having to change the color in Fig. 1.14) requires different modes and therefore implies a more fundamental re-design.

Huge leaps from one frequency to another can only be achieved by significant increase or reduction of the magnetic field strength in the cavity. It is evident that those leaps require at least the corresponding ramp-time of the magnet, which is typically of the order of tens of minutes, instead of seconds as for smaller frequency steps. A systematic compilation of various possibilities to change the frequency of a gyrotron is shown in Fig. 1.15.

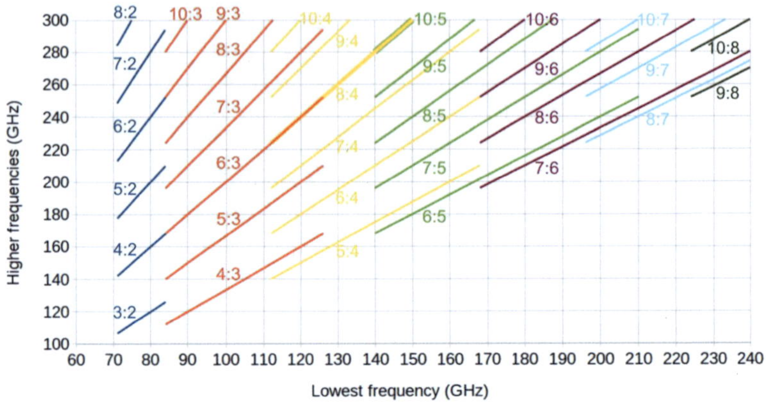

Figure 1.14: All possible frequencies for multi-frequency operation between 60 GHz and 300 GHz, according to the limitations described in the text. Same colors indicate same ratio between window resonance and lowest frequency (dark blue: 2; red: 3, yellow: 4; ...).

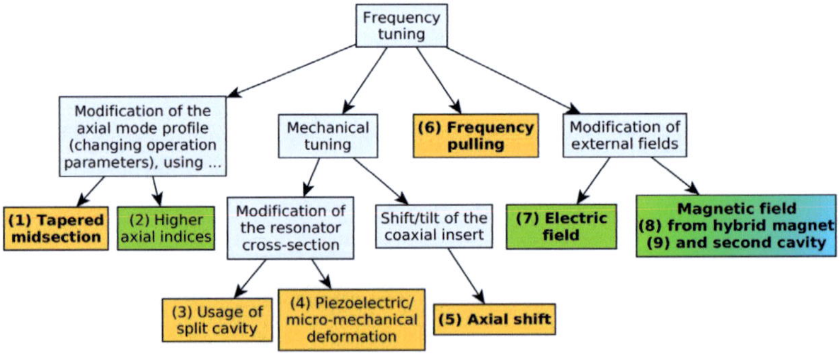

Figure 1.15: Frequency tuning possibilities as described in the text. Continuous and/or small-range ($\approx$ 1 GHz) finetuning methods are shown in orange; (quasi-)discrete, medium-range ($<$ 10 GHz) step-tuning possibilities in green; and operation at multiple, discrete frequencies over a wide range ($\approx$ 60 GHz) in blue. Techniques suitable for efficient high-power high-frequency gyrotrons are written in boldface. (1) [BNP76, GLMK08], (2) [HBG$^+$05], (3) [BCD$^+$84], (4) [SI08], (5) [IDT08, DM98], (6) [KDW$^+$84], (7) [Edg93], (8) [KDK$^+$01, Edg93], (9) [DP90].

1.3 Systematic Evaluation of Relevant Gyrotron Designs

1.3.1 State-of-the-Art Concepts

Existing coaxial-cavity and/or multi-frequency gyrotron designs serve as a technological baseline for this work. In this sense the following designs are most relevant:

1. The 1.5 MW 140 GHz coaxial-cavity gyrotron with a d-CIMIG and two output windows [FKM+94, PBD+98]. Its co-rotating operating mode $TE_{28,16}$ was intentionally converted to a counterrotating $TE_{76,2}$ mode before conversion into two $TEM_{0,0}$ beams and coupling-out of both through separate diamond windows. The first concept of this gyrotron had an axial RF output with a fused silica window.

2. The 1.5 MW 165 GHz coaxial-cavity gyrotron [PBD+97, PBD+99], which has been investigated in a large number of experiments.

3. The EGYC 2 MW 170 GHz coaxial-cavity gyrotron for ITER as described in [RPK+10]. A short-pulse preprototype of the gyrotron has been built and is used at KIT for study and improvement of the design.

4. Inspired by the considerations on the 2 MW design, a 4 MW design has been studied at IHM [Ber11]. Since present-day CVD-diamond windows are capable of transmitting at most 2 MW of radiation, this design has a dual-beam QOL and two windows. Several aspects of the design, for example multi-mode simulation results with more realistic beam parameters, remain for future investigation.

5. The multi-frequency gyrotron [SKO+13] developed at JAEA. This is a megawatt-class gyrotron that can be operated at four equally spaced frequencies from 104 GHz to 203 GHz, as determined by the window resonance frequency.

6. The coaxial $\mathrm{TE}_{42,7}$ mode cavity as proposed by Barroso and Correa [BC92], with a CW output power of 1 MW at 280 GHz. Their considerations did not lead to further detailed design studies. Other proposals for MW-class gyrotrons operating at the first harmonic with frequencies significantly above 200 GHz have been made recently by Kartikeyan *et al.*, e.g. [KJT13].
7. An approach towards step-tunable gyrotrons was described by Zapevalov *et al.* [ZBD+04]. The modes were chosen according to the limited possibilities of the diode MIG.
8. The dual-frequency TCV gyrotron with intended operating modes $\mathrm{TE}_{17,5}$ (84 GHz) and $\mathrm{TE}_{26,7}$ (126 GHz) [ABG+15], to which the mode-selection scheme given in Section 2.1.3 applies. However, according to that scheme, the upper mode should be $\mathrm{TE}_{25,7}$, due to the more similar caustic radius.

The design presented in this work is not based on any of the above designs in particular.

In order to quantify the complexity of individual gyrotron designs, especially of coaxial ones, one can take the mode eigenvalue as a measure. Design 3, for example, has a mode eigenvalue of $\chi_{34,19} = 105.2$ and its development is ongoing [RRP+16], while design 4 has $\chi_{52,31} = 168.6$ and is not under further consideration at the moment. Consequently, an indicator for the risk of manufacturing or operation failure could be defined as:

- $\chi \sim 100$: “low risk”
- $\chi \sim 150$: “medium risk”
- $\chi \sim 200$: “high risk”

If one restricts designs to such with medium risk and output frequencies around 240 GHz ($\lambda \approx 1.25$ mm), the target cavity radius is around 30 mm, close to that of design 2.

The window resonance frequency of modern gyrotrons for fusion applications is of course determined by their design frequency, which is 140 GHz for W7-X and 170 GHz for ITER, determined by the respective resonance heating requirements. Together with Fig. 1.14, only the following assignments of window resonances can be made:

- **140 GHz = 5 · 28 GHz:** The gyrotron described in [ZBD+04] falls into this category. It has the operating modes $TE_{19,6}$ (111.5 GHz), $TE_{22,8}$ (140 GHz) and $TE_{25,10}$ (168.3 GHz). Hypothetical higher frequencies of gyrotrons equipped with this window would be around 196 GHz, 224 GHz, 252 GHz, and 280 GHz.

- **170 GHz = 6 · 28.3 GHz:** This is almost the same design as above. Higher frequencies would be 198.3 GHz, 226.7 GHz, 255 GHz, and 283.3 GHz.

- **170 GHz = 5 · 34 GHz:** Having the design mode $TE_{31,11}$, the gyrotron described in [SKO+13] also operates at 104 GHz ($TE_{19,7}$), 137 GHz ($TE_{25,9}$) and 203 GHz ($TE_{37,13}$). A possible extension of this mode series to 238 GHz ($TE_{43,15}$) and 272 GHz ($TE_{49,17}$) has been considered in [KJT13] and [KAF+15b].

- **140 GHz = 4 · 35 GHz:** This variant was also discussed in [ZBD+04] (again with $TE_{22,8}$ as design mode, and $TE_{17,6}$ for 105.1 GHz) and [TAA+07] (design mode $TE_{28,8}$ and $TE_{21,6}$/$TE_{22,6}$ for around 105 GHz). The corresponding higher frequencies are 175 GHz, 210 GHz, 245 GHz, and 280 GHz.

One can see that the region around 240 GHz corresponds best to either 7 · 34 GHz or to 8 · 30 GHz.

1.3.2 Advanced Concepts

In order to enhance the gyrotron performance, one could consider gyrotron designs deviating from the generic designs [Thu16]. The following concepts have been considered in the frame of this work; however, none of them appeared to have convincing advantages over the generic designs and therefore a deeper analysis of these concepts has not been made.

1. **Fixed non-axisymmetric cavity:** Such a cavity would lead to different standing mode patterns in the cross-section, thus maybe reduce mode competition. On the other side, these geometries could lead to uneven distribution of the ohmic loading at the wall surfaces, would require an asymmetric beam shape (or suboptimum coupling), and would lead to a more complicated manufacturing process. Thus, such geometries are not practical. Only in the context of frequency tuning, Sabchevski and Idehara [SI08], and Brand *et al.* [BCD+84] have investigated non-circular cavities.

2. **Multiple concentric electron beams:** Those n beams could either be placed at the first n maxima of the operating mode to deliver power to it, or placed at the maxima of competing modes to extract power from them [ZT90]. Theoretical investigations of dual-beam gyrotron interaction [LYF+07] and of conceptual designs for dual-beam MIGs have been undertaken [FYYL09]. It has been claimed that two (or more) beams would reduce mode competition and enhance output power. However, the total output power of CW gyrotrons is limited by the ohmic losses on the outer wall, which results purely from the power transferred to the microwave, and by the cooling capacity of the collector, which has to collect electrons from all beams. Therefore it is obvious that, under otherwise unchanged conditions and limitations, one can only distribute the limited single-beam power over the n beams: multiple-beam gyrotrons are, in fact, "split-beam" gyrotrons;

and dual-beam designs cannot *per se* allow higher output power. Their primary advantage is that they would lead to a larger effective emitter width (since the individual widths add up), allowing smaller emitter radii and thus smaller MIGs. Disadvantages of such designs are the following:

- While beams, placed at second or higher radial maxima of the main mode, indeed decrease the coupling of the usual competitors, the coupling to many modes with lower h is increased, thereby creating new competitors.
- Higher radial maxima of the main mode have a smaller local field strength and therefore couple weaker to the electron beam. Therefore, the resulting coupling generally decreases with increasing number of beams.
- Already in single-beam gyrotrons, some of the processes within the beam and MIG region are not well understood. Having multiple beams would complicate things drastically like generation of secondary electrons, additional potential wells, possible beam-beam interactions, mutual misalignments, and so on.
- From a collector design point-of-view, multiple concentric particle beams are equivalent to one very thick beam (with gaps inside), which is a disadvantage according to present understanding, especially for multi-stage depressed collectors.

For gyrotrons unrelated to fusion plasma heating and control those disadvantages might be irrelevant. A recent example is the 780 GHz $TE_{8,5}$ second-harmonic double-beam gyrotron that is investigated in [GGM$^+$16], where the main purpose of the second beam is to avoid competition from fundamental-harmonic modes. The simulated beam voltage and total current are 20 kV and 2 A, respectively, and the electronic efficiency does not exceed 2.5 %.

3. **Elaborate cavity profiles:** Several proposals for elaborate longitudinal cavity profiles can be found in the literature, for example tapered cavities (e.g. [BNP76]), coupled or step-cavities (e.g. [ZMPT84]), cavities with output step, and cavities with arbitrary shape [BCM89]. With the design developed in this work (see Chapter 2), it has been confirmed that a small step – a fraction of a wavelength – between straight section of the cavity and uptaper indeed inceases the efficiency slightly, which is due to a changed axial mode profile. However, no systematic investigations into this were made, as this effect is not a specific coaxial-cavity property. Research on step-cavity gyrotrons has basically been discontinued around 1990. A systematic analysis of more elaborate profiles is beyond the scope of this thesis.

4. **Higher gyrotron harmonics:** As mentioned in Annex A.9, gyrotrons (including LOGs) can also be operated at higher harmonics s of the cyclotron resonance. However, it is well-known that the maximum possible interaction efficiency and hence the typical practical efficiency decreases with increasing s [Thu16]. Thus, for gyrotrons intended for nuclear fusion applications (high power, high frequency, high efficiency), only $s = 1$ seems reasonable, at least as long as single-stage depressed collectors are used.

5. **Tapered magnetic field:** In the past there were studies to use a tapered magnetic field in the cavity region, i.e. to place the field maximum at either end of the cavity, with the goal to increase efficiency or to decrease mode competition. However, both experimental [GB90] and theoretical [DK95] investigations showed that a magnetic field with its maximum at the cavity center, i.e. with minimum taper, is the optimum choice.

6. **Outer-wall corrugations:** There are two concepts behind the idea to introduce longitudinal corrugations on the cavity wall to decrease

mode competition, additionally to or alongside with corrugations on the insert. The first concept, termed “rippled-wall corrugations”, is to couple the azimuthal neighbors of the main mode to low-order modes with low quality factor (as indicated in Table 1.1) [CD84, Ker96]. Here, the corrugations are not smaller than the design mode’s wavelength, and the actual number of grooves determines which of the modes are coupled to each other.

The second concept is to use the outer wall as a second impedance boundary, as is under study by researchers in China [HYL15] and India [SJB04]. However, to the best knowledge of the author, it has never been demonstrated that the additional corrugations have any effect – whether recognizable or not – on interaction stability or other gyrotron properties; especially an effect that would justify the additional manufacturing (and cooling) effort.

2 Physical Design of Gyrotron Key Components

2.1 Basic Considerations

2.1.1 General Design Approach

The general goal of this work is to investigate the physical and technical feasibility of coaxial-cavity megawatt-class gyrotrons in a frequency range around 240 GHz. This goal has been achieved by proposing the physical design of a 2 MW 237.5 GHz coaxial-cavity gyrotron, described in this Chapter, and by investigating its critical properties in detail, as shown in Chapter 3.

For given caustic radius and material properties, ohmic losses on the cavity wall increase with increasing output power and/or frequency (see Eq. B.34). In order to keep the ohmic loading on the cavity wall at reasonable values, one therefore needs to increase the cavity radius, which requires modes of higher order. Such modes show stronger mode competition, making operation of the gyrotron less reliable. Thus, otherwise comparable gyrotron designs become more critical with increasing power and frequency; and conversely, the feasibility of gyrotrons with lower frequencies and/or lower output power is demonstrated by showing feasibility of higher-power higher-frequency designs.

In order to ensure that the design proposed in this work is a realistic one, all known possible issues arising in key components of the gyrotron have been identified and analyzed, as will be shown. A special focus has also been on the applicability of the used simulation codes in very high-order and high-frequency regimes.

The following additional **design goals** for the gyrotron were envisaged:

- Coaxial-cavity design to reduce mode competition (and, as additional features, to reduce voltage depression and to suppress parasitic HF oscillations in the beam tunnel).
- Fast frequency-step tunability.
- Multi-frequency operation (slow frequency tuning for multi-purpose operation).
- Total efficiency $\eta_{\mathrm{tot}} \geq 60\ \%$ at an output power as high as possible, at best significantly larger than 1 MW. Since this seems to be achievable only with Multi-Stage Depressed Collectors (MSDCs) which are not yet available, this work focuses instead on interaction efficiencies larger than 30 %.

The **design process** itself has been carried out in the following order, as has been shown useful for past gyrotron designs:

1. Choice of an appropriate design mode.
2. Cavity design and optimization.
3. Magnet design.
4. MIG design and optimization.

Since the designs of a QOL and of a single-stage collector for this gyrotron design are not expected to lead to scientific insight, they are not part of this thesis.

The following **technical boundary conditions** were respected, in accordance with [ADK^{+}08]:

1. Peak ohmic loading on the (outer) cavity wall $w_{\Omega,\mathrm{O}} \leq 2.0\ \mathrm{kW/cm^2}$, see Annex B.7. The peak loading is limited by the technically possible cooling capacity, which depends on coolant, pumping speed, and geometry of the cooling channels. Insufficient cooling would deform or eventually melt the cavity wall in long-pulse (CW) operation of gyrotrons. An electric conductivity for copper of $\sigma = 1.4 \cdot 10^7\ \mathrm{S/m}$ is considered, which already accounts for surface roughness and temperature of the heated cavity wall. Both values are relatively conservative [KAF^{+}16d] see also Section 2.5.2.

2. Peak ohmic loading on the insert wall $w_{\Omega,\mathrm{I}} \leq 0.2\ \mathrm{kW/cm^2}$. This value comes from technical considerations of the European 2 MW coaxial-cavity ITER gyrotron design. Values of $0.15\ \mathrm{kW/cm^2}$ and $0.1\ \mathrm{kW/cm^2}$ have been considered for the 1.5 MW, 165 GHz [PBD^{+}99] and 170 GHz [Rze07] tubes, respectively. In both cases, the ratio between permitted loadings on outer wall and insert is around 10, essentially due to the fact that the coolant has to flow both forth and back within the thin metallic rod.

3. Electron guiding-center spread ("beam thickness") in the cavity of $\varsigma_{r_\mathrm{b}} \gtrsim \lambda/5$. Values between $\lambda/8$ and $\lambda/4$ can be found in the literature. In the present design, the value of $\lambda/5$ for 170 GHz, considered safe, has been extrapolated to 240 GHz taking into account the scaling of the first maximum of the design mode. This led to $\varsigma_{r_\mathrm{b}} = \lambda/4.4$, or, equivalently, a single-side spread of $\zeta_{r_\mathrm{b}} = 1.4\ \%$ with respect to the mean guiding center radius.

4. Emitter radius $R_\mathrm{E} < 70\ \mathrm{mm}$. This target has been set to avoid a too large MIG diameter and thus a large bore-hole diameter of the gyrotron magnet, and it proved consistent with initial considerations

on basic emitter properties and a beam thickness of $\lambda/5$. No explicit restrictions concerning the magnetic compression have been imposed on the design.

5. Electric field in the MIG region $E_{\mathrm{G,max}} < 7\ \mathrm{kV/mm}$, especially at the emitter, to avoid arcing (Penning discharges).

6. Maximum current density from the emitter surface $j_{\mathrm{E}} \leq 4\ \mathrm{A/cm^2}$ to avoid fast aging of the emitter (see Fig. 1.9).

7. Radial quasi-optical output coupler (launcher), suitable for single-beam output.

8. Single-disk CVD-diamond output window.

Since a coaxial-cavity design is considered, voltage depression on the electron beam (see Annex A.4) and accompanying effects such as limiting current are not considered in the design phase. Limitations on anode-cathode distance (with respect to gyration radius), space-charge-limited current density, diffractive quality factor with respect to its minimum value, percentage of ohmic losses, and Fresnel parameter were not directly taken into account during the design phase, since their values are by definition acceptable if the simulations return reasonable results.

The **assessment of technical feasibility** includes:

- Tolerance studies of cavity and MIG.
- Discussion of possible criticalness of other gyrotron components (e.g. beam tunnel, collector).

2.1.2 Definition of the Design Process

Formulas in gyrotron textbooks [Edg93, KBT04, Nus04] usually interrelate design parameters of MIG, cavity, and launcher, but leave it up to the reader to decide which of these parameters are considered as the independent and which as the dependent variables. During the design process of a gyrotron, this causal relation is determined by the imposed restrictions and requirements, at least to a significant degree.

The dependence relations of parameters for one possible design approach is shown in Fig. 2.1. Externally imposed and basically independent parameters are shown in green and red boxes, where green parameters are design requirements (here: a certain frequency, output power, and multi-frequency operation) and red parameters are physical or technical restrictions. Derived parameters are given in grey, yellow, and cyan boxes, where cyan and yellow refer to the parameters that fix a design (e.g. operating mode, geometry, voltages). Of these, parameters in yellow boxes are usually restricted to some extent.

One can especially see that beam parameters following from the cavity design, for example beam current or pitch factor, are at a later stage related to MIG parameters, for example emitter current density and emitter slant angle, which themselves are restricted. Cavity parameters can furthermore be related to launcher and window requirements such as mode indices allowed in multi-frequency operation. Thus, a proper cavity design process has to respect MIG and launcher constraints. It is hence obvious that a serious gyrotron design requires a complex determination and evaluation of all these parameters if one tries to optimize the design without violating the given physical and technical restrictions, at best some sort of automatic algorithm. While codes existed which determine, for example, the maximum possible output power for some given boundary conditions (at IHM: MAXPO of the CAVITY package, see Annex C.2.1), the more detailed "zero-dimensional" design code Design-o-mat has been written in the frame of this work (see Annex C.2.3) to implement the processes shown in Fig. 2.1.

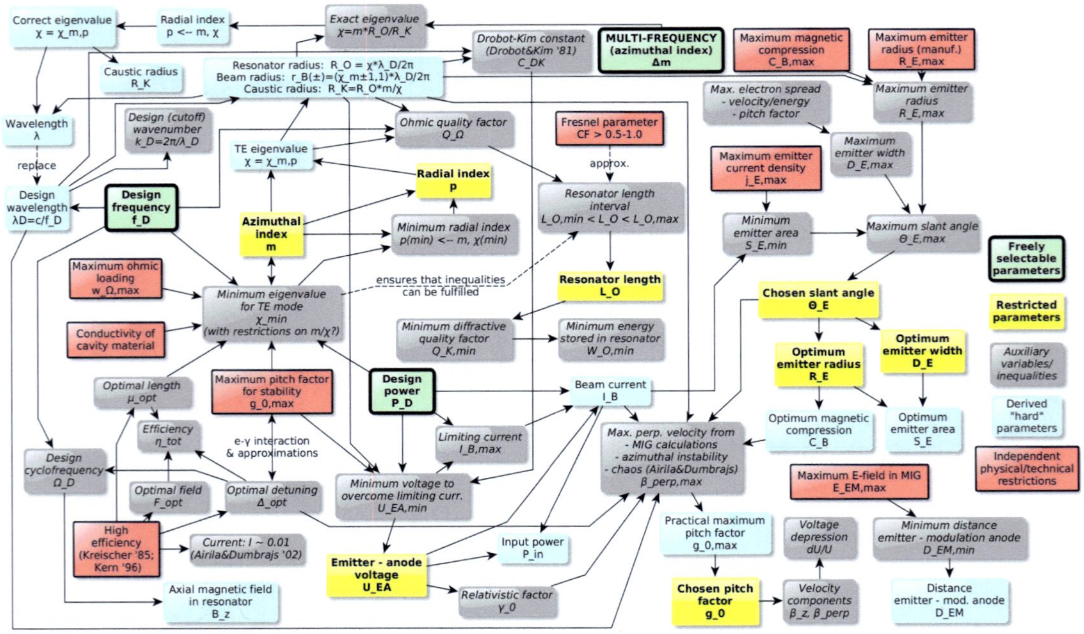

Figure 2.1: Flowchart of a possible approach towards the design of MIG and cavity of a gyrotron. Design parameters which can be freely chosen are given in green boxes; parameters which are restricted to certain values in yellow; independent physical or technical restrictions in red; dependent, auxiliary variables in grey; derived variables, which are important design parameters, in cyan.

2.1.3 Novel Systematic Mode Selection Strategies

Mode Selection Based on the Multi-Frequency Requirement

In [FIA$^+$14a, FAG$^+$15], the theoretical basis behind a novel mode-selection strategy has been published which respects the requirement for multi-frequency operation. It is demonstrated there that very high-order modes have eigenvalues which are basically linear in their azimuthal and radial indices (see B.9), which leads to coupling spectra that are almost identical for modes with similar indices, see Fig. 2.2. Since spectral features around, and axial and transverse profiles of such modes are also very simular, parameters such as output power and efficiency of the respective gyrotron designs will depend much less on the particular chosen mode than on other factors.

Therefore, one may choose the operating modes according to other criteria, for example multi-frequency operation. This mode-selection criterion is solely based on the assumptions that the gyrotron in question has

- a quasi-optical launcher, and
- a single-disk window with realistic thickness,

which is common for state-of-the-art gyrotrons. As already explained, while the window requires operating frequencies that are integer multiples of its resonant frequency, the QOL requires modes with similar relative caustic radii m/χ. None of those criteria can be fullfilled exactly by any set of modes, but one can define a combined figure of merit that helps to find the most suitable mode series. The figure of merit – termed "rating" – as defined in [FAG$^+$15] returned fifteen mode series that show a very good multi-frequency behavior. Two of these series are:

- $\mathrm{TE}_{28,17}$, $\mathrm{TE}_{35,21}$, $\mathrm{TE}_{42,25}$, $\mathrm{TE}_{49,29}$, $\mathrm{TE}_{56,33}$
(series (b) in [FAG$^+$15])
- $\mathrm{TE}_{28,13}$, $\mathrm{TE}_{35,16}$, $\mathrm{TE}_{42,19}$, $\mathrm{TE}_{49,22}$, $\mathrm{TE}_{56,25}$
(series (e) in [FAG$^+$15])

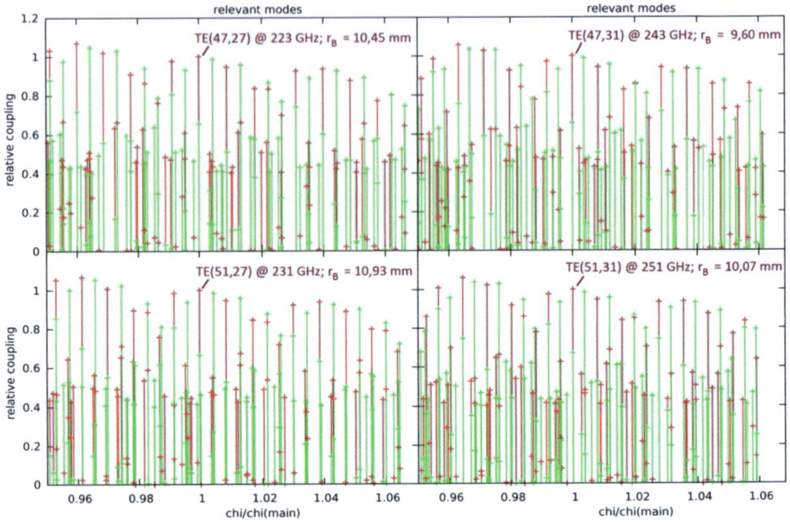

Figure 2.2: Coupling spectra of four modes around $\mathrm{TE}_{49,29}$, the beam radius each time being chosen for maximum coupling (see Eq. 1.19). Coupling strengths and eigenvalues are normalized to the respective value of the main mode. Co-rotating modes are shown in red, counterrotating modes in green.

For both series, the detailed properties of the three modes in the middle are given in Table 2.1. Frequencies have been chosen such that the lowest mode corresponds to exactly 170 GHz – which is the only reason why the design frequency of the gyrotron in this work is not exactly 240.0 GHz –, while the highest mode corresponds exactly to the seventh reflection minimum of the window ($D_\mathrm{W} = 7 \cdot \lambda_{49,p}/2$). The corresponding window thickness is 1.857 mm (see Section 1.2.3). The output frequency of 170 GHz could be used during the startup phase in DEMO (or ECRH in ITER), while 204 GHz and 238 GHz can be used for bulk heating and ECCD in DEMO, respectively.

Table 2.1: Basic properties of the modes of series (b) and (e) from [FAG+15].

Series (b) mode	$TE_{35,21}$	$TE_{42,25}$	$TE_{49,29}$
Eigenvalue	113.1	135.6	158.1
Ideal cutoff frequency (GHz)	**170.00**	203.75	**237.50**
Window resonance (GHz)	169.64	203.57	**237.50**
Relative caustic radius	0.3094	0.3097	0.3100
Series (e) mode	$TE_{35,16}$	$TE_{42,19}$	$TE_{49,22}$
Eigenvalue	96.5	115.6	134.7
Ideal cutoff frequency (GHz)	**170.00**	203.71	**237.41**
Window resonance (GHz)	169.58	203.50	**237.41**
Relative caustic radius	0.3628	0.3634	0.3637

Mode Selection Based on Mode Competition

Coaxial-cavity gyrotrons tend to use operating modes with higher h_{mp}, i.e. volume modes, in order to make efficient use of the higher possible beam power at a given cooling capacity of the cavity wall and of the reduced voltage depression (Annex A.4). It has already been pointed out (Table 1.1) that the main competitors of co-rotating mode TE^{co}_{mp} are the counterrotating modes with smaller caustic radius, $TE^{ct}_{(m-2),(p+1)}$ and $TE^{ct}_{(m-3),(p+1)}$. An evaluation of four existing coaxial gyrotron designs [FKM+94, PBD+99, ADK+08, Ber11] shows that the operating modes lie on the line $p : m \approx 0.57$ (or, equivalently, $C_K \approx 0.31$), regardless of their output power or frequency, as has been presented on several occasions [Fra13a, Fra13b, FAI13], see Fig. 2.3. This is the condition where the main mode is equally separated from its two main competitors; therefore, one can regard that condition as an additional mode-selection criterion for coaxial-cavity designs. The combination of both mode-selection strategies results in series (b) with design mode $\mathrm{TE}^{\mathrm{co}}_{49,29}$ having an eigenvalue of $\chi \approx 158.058$ as the optimum choice. Up to this eigenvalue, around 3200 modes exist, compared to e.g. 640 for the mode $\mathrm{TE}_{32,9}$ mentioned in [ADK+08]. The mode $\mathrm{TE}_{35,21}$ ($\chi_{35,21} = 113.133$), corresponding to 170.0 GHz, has only slightly higher mode indices than the one considered for the coaxial EU ITER gyrotron ($\mathrm{TE}_{34,19}$, $\chi_{34,19} = 105.193$) [RPK+10].

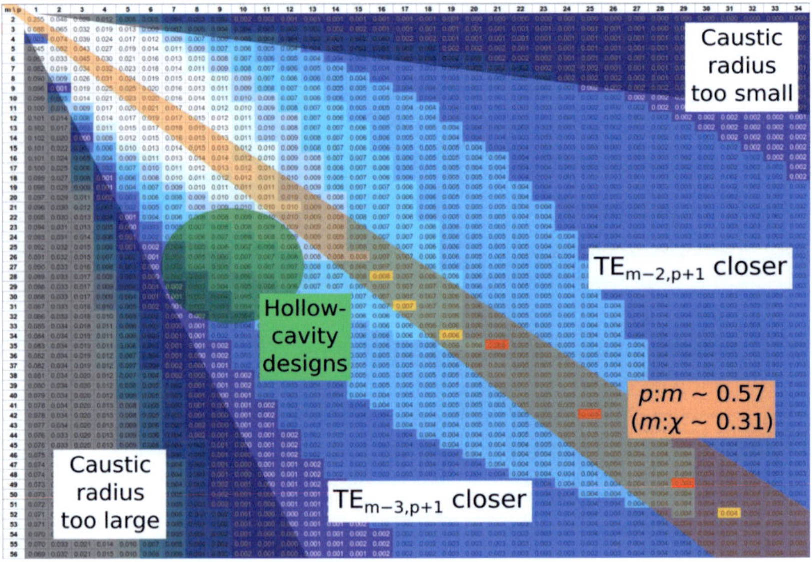

Figure 2.3: Spectral distance of co-rotating modes to their closest counterrotating competitor in the m-p plane. The shades of blue represent the relative distance to the respective competitor $\Delta\chi/\chi$, where darker shades indicate smaller distances. Also indicated are areas of very small and of very large caustic radius (in which mode selection is not relevant), the areas of typical hollow-cavity (green; see [Thu16]) and of coaxial-cavity (orange; see text) modes and which competitor is the closer one. The modes from [FKM+94, PBD+99, ADK+08, Ber11] are marked yellow and the mode series [FAG+15] (b) proposed in this work in red.

There have been studies conducted on whether there is a maximum possible operating mode, at least for given boundary conditions, and what the indices of such a mode are. Generally, studies suggest that there might be problems above $m = 45$ [AD02, DKR05]. However in the frame of this thesis, a $\mathrm{TE}_{49,29}$ gyrotron design with parameters very close to the present design has been investigated for possible azimuthal instabilities [DAFJ14], and it was concluded that azimuthal instability does not necessarily occur, especially for high beam current.

2.2 Axisymmetric Coaxial Cavity Design

The CAVITY code package (routines GEOMT, COLDC, MCONV; see Section C.2.1) has been used to find a cavity design for a coaxial $\mathrm{TE}_{49,29}$ ($\chi_{49,29} = 158.058$) mode gyrotron with 237.5 GHz as the operating frequency. For initial studies a hollow cavity was considered. Coupling spectra of the modes to the electron beam were performed using SCNCHI. A counterrotating beam has been used for some startup scenarios, but the gyrotron performance was not as good as with a co-rotating beam. The optimized geometry for the $\mathrm{TE}_{49,29}$ mode cavity is given in Table 2.2 (first presented in [FIA$^+$14b]). The exact cavity radius for a standing wave would be around 31.754 mm, but to obtain a slightly forward-directed wave the cavity radius has been chosen 31.780 mm, or 0.083 % larger (corresponding to cutoff frequency 237.30 GHz). Parabolic smoothings were implemented to connect the midsection with the downtaper and uptaper sections. See Fig. 2.4 for an illustration of the final cavity design (including the coaxial insert, but without nonlinear uptaper). The midsection part of the cavity wall has no taper.

The CAVITY routine GYMOT was then used to find initial values for the beam parameters (current, voltage, pitch factor) and for the magnetic field to be applied. These values were refined after obtaining self-consistent solutions with SELFC. In order to find an initial operating point and an appropriate start-up scenario, SELFT was used.

Table 2.2: Geometric parameters of the cavity.

Electron beam radius	r_b	(10.24 ± 0.143) mm
Cavity radius	R_O	31.78 mm
Downtaper length	$L_{\mathrm{O},1}$	10 mm
Downtaper angle	$\Theta_{\mathrm{O},1}$	3.5°
Smoothing downtaper–midsection	$H_{\mathrm{O},1}$	2 mm
Midsection length	$L_{\mathrm{O}(,2)}$	15 mm
Smoothing midsection–uptaper	$H_{\mathrm{O},2}$	4 mm
Uptaper length	$L_{\mathrm{O},3}$	20 mm
Uptaper angle	$\Theta_{\mathrm{O},3}$	3.0°
Total length		45 mm

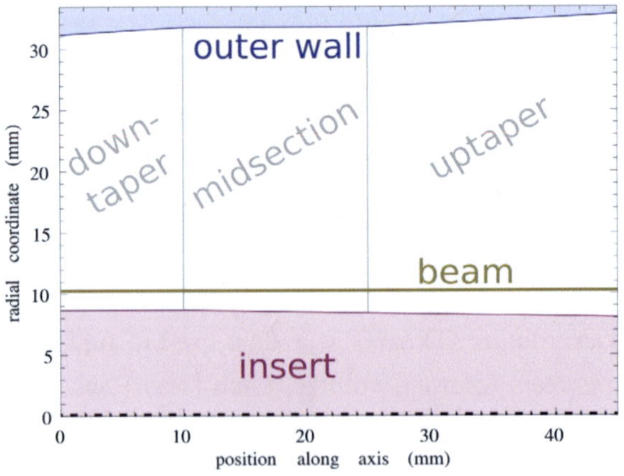

Figure 2.4: Cavity design for the 237.5 GHz, $\mathrm{TE}_{49,29}$ mode gyrotron. Axial and radial coordinates have the same scaling, in contrast to many cavity depictions, to emphasize the fact that the straight part of the cavity is shorter than 1/4 of its diameter. Shorter cavities are a general trend observed for most high-power, high-frequency gyrotrons.

During the studies it has been found that SELFT has severe limitations, especially that it is not possible to simulate a thick electron beam (guiding center spread), that ohmic loading is not calculated and that the adverse effect of electron velocity spread on gyrotron performance is exaggerated significantly. Therefore, COLDC, SELFC and SELFT were replaced by COLDCAV and EVRIDIKI from the EURIDICE package [APIV12, Avr15], see Annex C.2.2.

A second issue with the CAVITY package is the correct calculation and numbering of mode eigenvalues for coaxial cavities. In order to solve this, the affected routines SCNCHI and CHIMP were replaced by a completely new routine SCNCHIMP. See Fig. 2.5 for comparison between CAVITY and the design codes used in this work.

The geometric parameters of the coaxial insert resulting from simulations using the axisymmetric Surface Impedance Model (SIM) are given in Table 2.3, with groove parameter $s/t = 0.5$. (For information on the actual corrugations, see Chapter 3.) The optimum operating parameters (angles optimized in steps of 0.5 degrees, lengths optimized in steps of 1 mm) and results according to self-consistent, time-dependent multimode simulations [FAI+14] with EVRIDIKI are given in Table 2.4. The distance between (nominal) beam radius and (nominal) insert radius, expressed in Larmor radii, is $(10.24 - 8.66)/0.08 = 20$, which is the same, safe value as in the coaxial 2 MW ITER gyrotron [RPK+10]. In order to ensure frequency step-tunability, a magnetic field of up to 10 T is needed.

The alternative main operating mode $TE_{49,22}$, having a larger caustic radius and thus resembling more a hollow-cavity mode, has also been studied in some detail. Cavity geometry (derived from the $TE_{49,29}$ mode design) and operating parameters are given in Table 2.5.

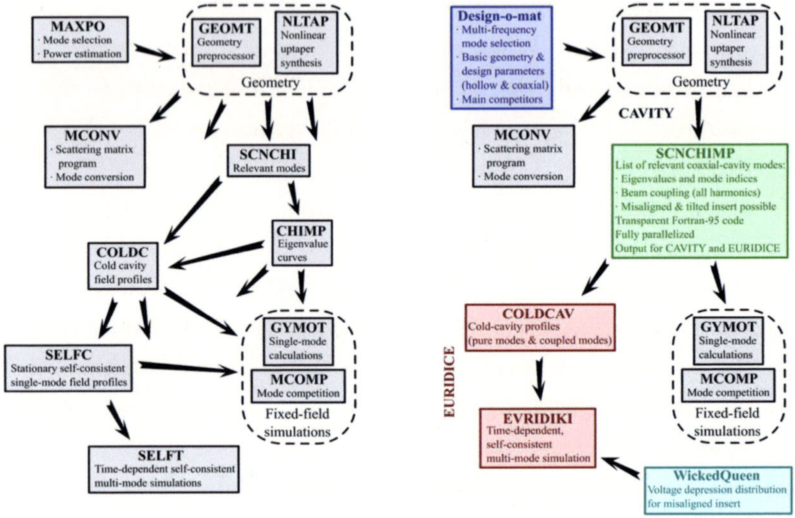

Figure 2.5: The CAVITY package as of end 2012 (left; [Ker96], own translation) compared to the cavity design and interaction codes as presented in the frame of this work (right).

Table 2.3: Geometric parameters of the insert. The corrugation depth given is the parameter used for SIM-based modelling and is not necessarily equal to the depth of realistic corrugations, depending on their shape.

Reference radius (in upstream part)	$R_{\mathrm{I},1}$	$8.66\ \mathrm{mm}$
Upstream part length	$L_{\mathrm{I},1}$	$10\ \mathrm{mm}$
Upstream part angle	$\Theta_{\mathrm{I},1}$	$0.0°$
Width of parabolic smoothing	$H_{\mathrm{I},1}$	$2\ \mathrm{mm}$
Downstream part length	$L_{\mathrm{I},2+3}$	$(15+20)\ \mathrm{mm}$
Downstream part angle	$\Theta_{\mathrm{I},2+3}$	$-1.0°$
Corrugation depth	$D_{r,\mathrm{I}(,2)}$	$0.3\ \mathrm{mm}$

Table 2.4: Optimum operating parameters and simulation results for the $TE_{49,29}$ mode design.

Cavity magnetic field at nominal beam position	B_{O}	$9.58\ \mathrm{T}$
Electron energy at the cavity entry	W_{O}	$85.6\ \mathrm{keV}$
Voltage margin to mode loss (overshoot)	δU	$2\ \mathrm{kV}$
Beam current	I_{b}	$69.3\ \mathrm{A}$
Nominal pitch factor at cavity entry	g_{O}	1.22
RMS spread in perp. velocity component	$\zeta_{v_\perp}$	$6\ \%$
Peak ohmic loading on the cavity wall	$\hat{w}_{\mathrm{O}}$	$2.0\ \mathrm{kW/cm^2}$
Peak ohmic loading on the coaxial insert	$\hat{w}_{\mathrm{I}}$	$0.2\ \mathrm{kW/cm^2}$
Output power	P_{out}	$1.9\ \mathrm{MW}$
Electronic efficiency	η_{elec}	$33\ \%$

2.3 Electron-Optical System Design

In this work, the focus of investigations was on the upstream part of the electron-optical system, i.e. the magnetron injection gun, the compression zone in the beam tunnel region, and the cavity. Detailed considerations on the spent beam in the downstream region, especially the collector, have not been undertaken; for a general discussion, see Section 2.4.

2.3.1 Fundamental Magnetron Injection Gun Properties

As a first approach towards the geometry of the MIG, the emitter slant angle Θ_{E} has been fixed to $25°$ (angle between surface normal and gyrotron axis: $65°$), under the assumption that this value is a good compromise between producing a thin electron beam and making sure that the beam is laminar, i.e. without intersecting electron trajectories.

Table 2.5: Geometric and interaction simulation parameters of the $TE_{49,22}$ mode cavity (without consideration of velocity spread).

Electron beam radius	r_b	(10.18 ± 0.08) mm
Cavity radius	R_O	27.10 mm
Downtaper length	$L_{O,1}$	10 mm
Downtaper angle	$\Theta_{O,1}$	3.5°
Smoothing downtaper–midsection	$H_{O,1}$	2 mm
Midsection length	$L_{O(,2)}$	15 mm
Smoothing midsection–uptaper	$H_{O,2}$	4 mm
Uptaper length	$L_{O,3}$	20 mm
Uptaper angle	$\Theta_{O,3}$	3.0°
Total length		45 mm
Reference insert radius	R_I	ca. 8.55 mm
Cavity magnetic field	B_O	9.26 T
Electron energy at cavity entrance	W_O	63.8 keV
Beam current	I_O	59.3 A
Nominal pitch factor cavity entrance	g_O	1.24
Output power	P_{out}	ca. 1.5 MW

With an emitter radius R_E of 65 mm and an emitter width of 4.3 mm (3.9 mm projected to the z-axis), a beam current of 70 A can safely be generated; and the emitter width projected to the gyrotron's cross-section is 1.8 mm, which is slightly below 2.8 % of the emitter radius. During the detailed design process described in Section 2.3.3 it could be confirmed that these values were a good choice and did not have to be changed afterwards.

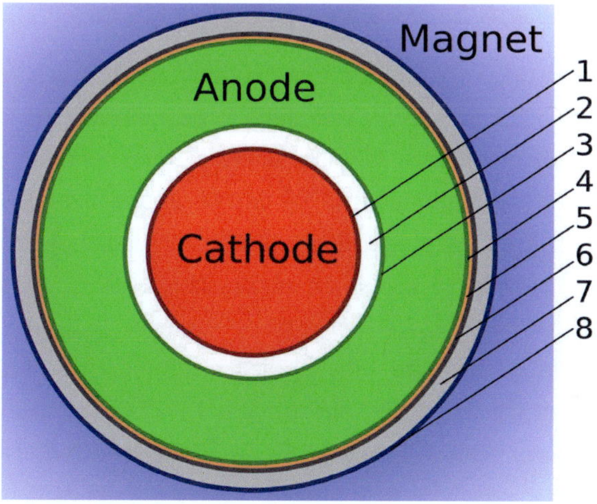

Figure 2.6: Radial dimensions of emitter, anode, and innermost parts of the superconducting magnet. (1) emitter radius at 65 mm, (2) vacuum, (3) outermost part of the vacuum at 80 mm, (4) outermost part of the anode at $\lesssim$ 135 mm, (5) high voltage oil, (6) bore-hole of the magnet at 135 mm (i.e. innermost part of the magnet), (7) thermal insulation within the magnet, (8) innermost part of the superconducting coil at 150 mm. See also Fig. 1.7.

In European high-power gyrotron designs, the radial distance between emitter and outmost wall of the MIG is around 75 mm to 80 mm [PBD+97, Rze07, PDD+05]. This distance is partly due to the required vacuum between cathode and (modulation) anode, see Fig. 1.10, and partly due to the structural stability of the outer electrode, its cooling channels, and the necessary ceramics between the outboard electrodes. In this work, a slight improvement has been assumed, reducing the required distance to 70 mm. Therefore, the present design would require a magnet bore-hole diameter of $D_{\mathrm{H}} = 2 \cdot (65\ \mathrm{mm} + 70\ \mathrm{mm}) = 27\ \mathrm{cm}$, see Fig. 2.6 for a visualization.

Although not considered in this work, an inverse MIG could lead to less material necessary for strucutral stability, cooling, and insulation – in addition to removed potential wells (see Section 1.2.2 and references therein) –, and thus to an even smaller required magnet bore-hole.

2.3.2 Requirements for the Superconducting Magnet

As discussed in Section 1.2.2 the gyrotron magnet field is a connectional entity between MIG and cavity; therefore, it makes sense to base the MIG design on a realistic magnet. The bore-hole diameter and the central field of the magnet are critical values of its design, and it had quickly become clear that no magnet design exists yet that fulfills the abovementioned requirements. Therefore, a survey on appropriate, realistic magnet designs was initiated.

In addition to the nominal values given in the previous Section, a certain flexibility of the magnetic field strength and direction at the emitter is desired to perform tests and experiments with the gyrotron. However, the inclination of the magnetic field should always remain zero in the cavity and be sufficiently constant around the interaction point. With four restrictions of the field (strength and angle at both emitter and cavity positions), the magnet needs at least four degrees of freedom, i.e. four independent coil connections. In anticipation of the discussions on the launcher design, Section 2.4, the magnet should not expand the electron beam too early such that it could hit parts of the launcher, see Fig. 1.7 for illustration.

In previous market surveys it was requested from the magnet manufacturers to reproduce the magnetic field profile that had been used in design simulations. However, in the frame of this study, it has been attempted to define the magnet requirements as loosely as possible to give potential manufacturers the most freedom in the magnet design. For further simplification, the requirements were formulated in terms of magnitude and gradient of the axial magnetic field (using Eq. A.8). The final values are listed in Table 2.6.

Requirements #1 and #2 refer to the magnetic field strength and homogeneity in the cavity, #3 refers to the field strength at the emitter (or, alternatively, to possible magnetic compressions), #4 refers to possible angles of the field at the emitter, and #5 refers to the maximum electron beam radius at the launcher. See Fig. 2.7 for a depiction of a magnetic field profile that

Table 2.6: Main requirements for the magnetic field (B-) profile along the gyrotron axis. See text for further explanation.

No.	Description of requirement	Value
#1	Minimum peak magnetic field (defining axial coordinate $z = 0$)	$B_{\mathrm{O,max}} = 10.5$ T
#2	Deviation from $B_{\mathrm{O,max}}$ in the region -5 mm $\leq z \leq +5$ mm	$\lvert\delta B\rvert \leq 0.001\ B_{\mathrm{O,max}}$
#3	Achievable field strengths B_{E} at one selected position z_{E} in the region -430 mm $\leq z_{\mathrm{E}} \leq -380$ mm (z_{E} by choice of the manufacturer)	$0.02\ B_{\mathrm{O,max}} \leq B_{\mathrm{E}} \leq 0.05\ B_{\mathrm{O,max}}$
#4	Achievable field gradients $\mathrm{d}B/\mathrm{d}z$ at the fixed position z_{E} selected in #3	-0.5 T/m $\leq \mathrm{d}B/\mathrm{d}z \leq 2$ T/m
#5	Minimum B-field at position $z = +400$ mm	$B \geq 0.12\ B_{\mathrm{O,max}}$
#6	Minimum bore-hole diameter D_{H}	$D_{\mathrm{H}} \geq 270$ mm

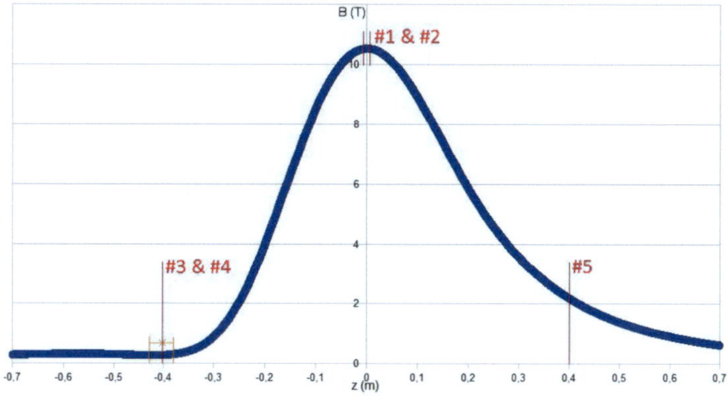

Figure 2.7: Main requirements for the magnetic field (B-) profile along the gyrotron axis. The regions listed in Table 2.6 are highlighted.

would fulfil these requirements. One design which met the requirements was obtained and used for the MIG design and subsequent studies.

2.3.3 Detailed MIG and Compression Zone Design

In this work the MIG design has been limited to the physical design and electron-optical analysis of the components around and downstream of the emitter. The rear parts of the MIG as well as technical features such as ceramic rings, cooling pipes, or emitter details were not considered. One reason for this is that the underlying magnet design is just an initial study and that several geometric features of the MIG depend heavily on the details of the field of the (final) magnet. Another reason is that it is apparently not necessary to find a completely new approach towards MIG designs; hence, no new findings are expected during a detailed design of technical components. Furthermore, only a rough basic geometry of the beam tunnel has been considered: The width of the ceramics ring between modulation anode and body might be too small, and the diaphragm at the end of the tunnel has a sharp edge to simulate a worst-case influence on the beam.

The basic physical t-CMIG design has been obtained using the code ARIADNE [PV04], which, in turn, has been extended and improved to cope with the t-CMIG geometry. One distinctive feature of the design is the modulation anode step which contributes significantly to both quality and compactness of the MIG. See Fig. 2.8 and [FAI$^+$14, FPA$^+$15c].

In order to avoid the phenomena described in Section 1.2.2 and references therein, the basic MIG design fulfills the following design criteria simultaneously:

1. Electrons starting from anywhere on the cathode surface at zero velocity either hit one of the anodes (modulation anode, body anode, or insert) or enter the cavity with low pitch factor.

2. Potential wells are avoided or at least reduced to low voltages such that trapped electrons need only a small kinetic energy to escape.

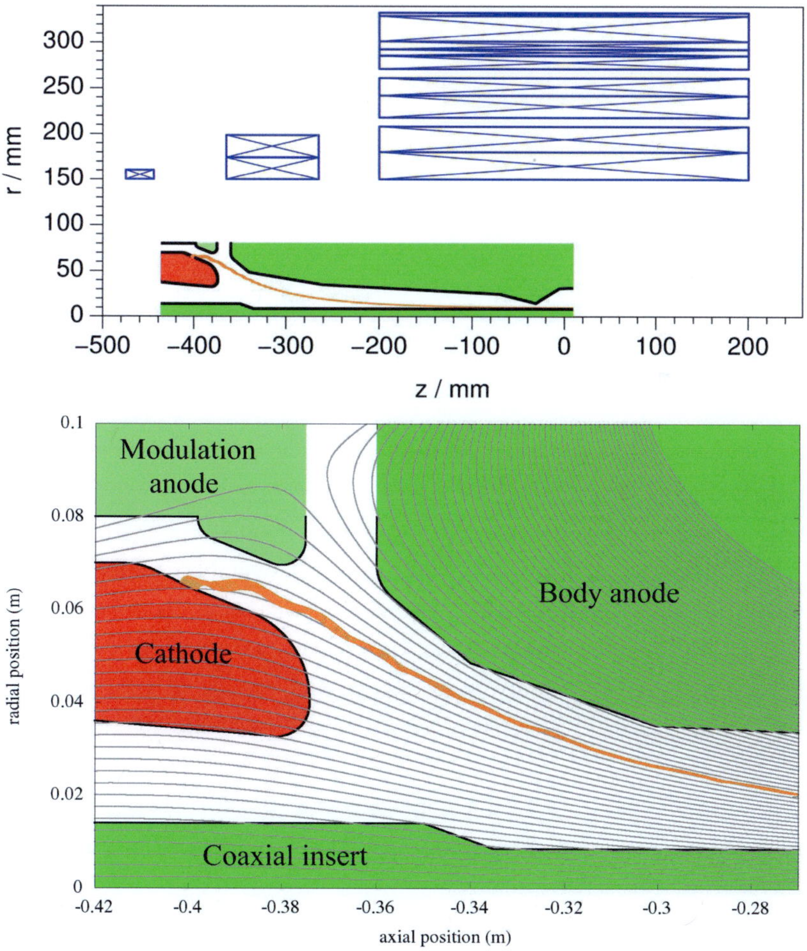

Figure 2.8: *Upper:* Basic design of the triode coaxial MIG, magnet coils, beam tunnel and cavity. *Lower:* Details of the MIG section with magnetic field lines (grey) and electron beam (orange).

3. The maximum electric field strength inside the MIG, especially at the emitter surface, is small enough such that arcing between electrodes is unlikely to happen.

4. The MIG is reasonably compact.

Figures 2.9 through 2.11 show that these three criteria are met: The pitch factors of beam electrons in the cavity are rather constant at 1.22, while none of the electrons emitted from other regions, i.e. any potential secondary electron, has a pitch factor above 2.5, which is considered a safe value [PPZ+16]. In full simulations no particles are reflected back from the cavity region. Two minor potential wells exist, one at the modulation anode step, and one in the beam tunnel. While the well in the beam tunnel can be reduced further during a detailed design, any attempt to reduce the potential well at the anode step resulted in a worse beam quality.

For verification and further investigations, the code ESRAY [IZJ15] was used. For several operating points, the resulting beam parameters (pitch factor, RMS spread of perpendicular component, and beam thickness) from both codes were compared. The results for pitch factor and beam thickness were always in good agreement, far within the limits of expected experimental accuracy.

The RMS spread of the perpendicular velocity component $\zeta_{v_\perp}$ was always two to three times higher in ARIADNE simulations than in ESRAY simulations. This is consistent with previous comparisons of the code. However, these spread values of 0.2 to 0.7 percent only occur if the ideal condition of a perfectly smooth emitter surface is assumed. ESRAY, in turn, can model emitter surface roughness, which influences the initial velocity distribution of the beam electrons [IZJ15, ZIP+16]. Using a typical structure size of 2 μm, the spread increases to 3 % or above, see Fig. 2.12 and Tables 2.8-2.9.

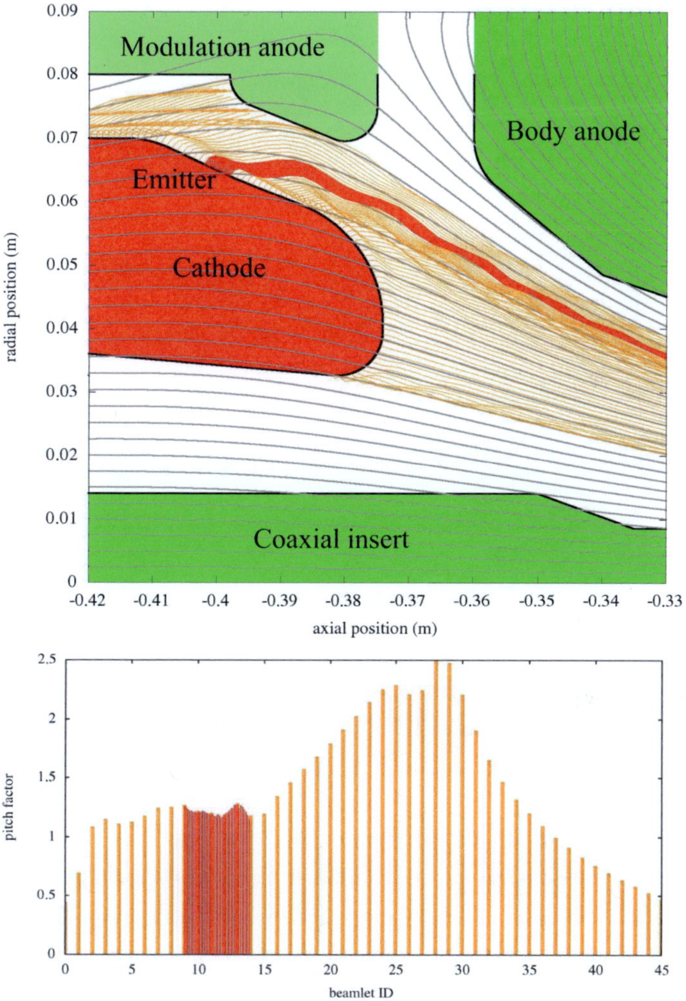

Figure 2.9: *Upper:* Trajectories of electrons from the emitter (red) and from other cathode surfaces, i.e. possible secondary electrons (orange), in the MIG region. Grey: magnetic field lines. *Lower:* Pitch factor distribution at the cavity entrance of those electrons which are not caught by the modulation anode or by the insert.

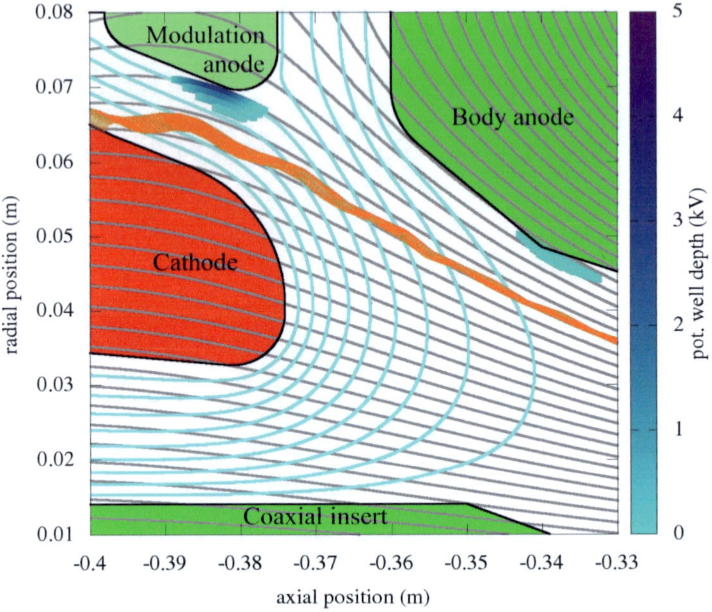

Figure 2.10: Visualization of potential wells in the MIG region. Shown are metallic surfaces (black), the electron beam (orange), magnetic field lines (grey), electric equipotential lines (cyan) and the depth of local potential wells. The potential wells cover only a small space and their depth is below 5 kV.

Azimuthal emission nonuniformity, caused by nonuniform heating or by a nonuniform work function, for example, has not been explicitly considered in this work. However, there is a margin left between the 3.4 % spread delivered by the MIG and the 6 % taken as input in the interaction simulations.

For nominal electrode voltages during $\mathrm{TE}_{49,29}$ mode operation see Table 2.7. The difference between accelerating voltage (87.43 kV) and beam voltage (85.6 kV) is due to the voltage depression caused by the beam space-charge. In the results presented here, no beam neutralization has been assumed. Calculations including beam neutralization showed, however, very similar results: basically, only the kinetic energy of the electrons changes

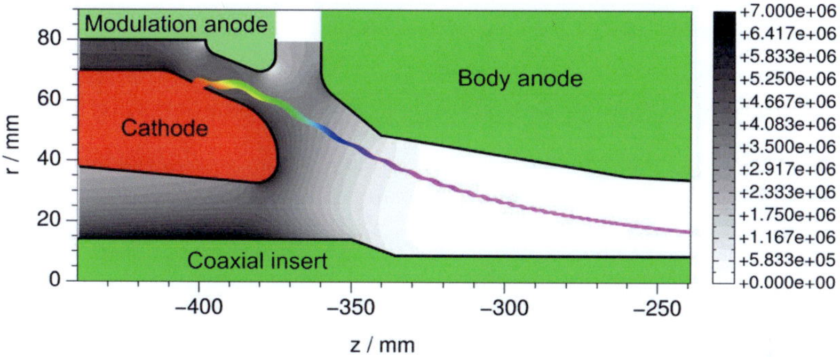

Figure 2.11: Electric field strength in V/m (grey shades) and electron beam (colored according to the local γ factor) in the MIG region. As one expects from basic electrostatics, the electric field is strongest where potential differences are large, where distances are small, and where roundings are significant. This is the case in both directions from the emitter, at the inner edge of the cathode, and at the insert surface where it lies within the cathode. The emitter region has been designed such that the beam crosses only areas where the electric fields are homogeneous, which decreased velocity spread.

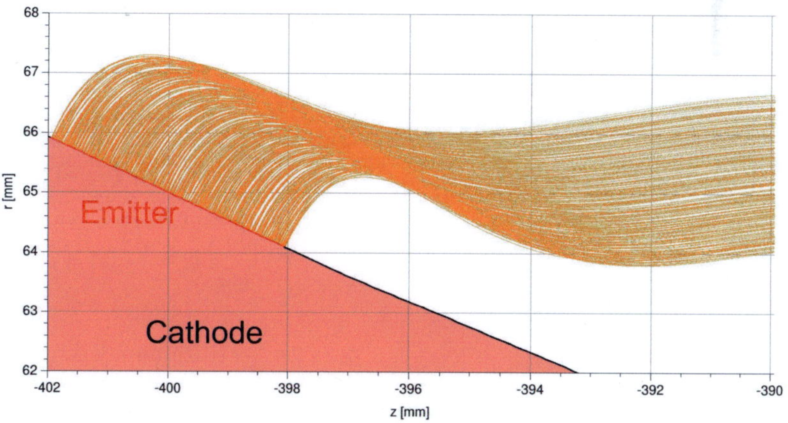

Figure 2.12: Trajectories of electrons released from an emitter (red) with rough surface (structure size: 2 μm), modeled by ESRAY. Geometry and parameters are the design parameters given in the text.

by the magnitude of the voltage depression. The anode potential has been set to ground potential, but of course, only the voltages between the electrodes matter. Figure 1.7 shows the typical electric connection of gyrotron components.

Table 2.7 also shows the resulting beam parameters in the cavity immediately before the interaction point, calculated with ESRAY. Clearly, the MIG delivers the beam required by the interaction calculations. Furthermore, the magnetic field strength at the emitter position is given. One can see that the magnetic compression between emitter and cavity as determined by the field strengths is almost 25 % higher than the value determined from the beam radii of 40, which is due to the nonadiabaticity of the electrons near the emitter.

Table 2.7: Nominal electrode potentials and beam parameters, including relative standard deviations from mean values. Statistics were taken from electrons within the last five millimeters before the cavity center, without consideration of any gyrotron interaction inside the cavity.

Anode & insert potential	$\Phi_{\mathrm{A}}, \Phi_{\mathrm{I}}$	0 kV
Cathode potential	Φ_{E}	-87.43 kV
Modulation anode potential	Φ_{M}	-37.45 kV
Electron kinetic energy	W_{O}	85.61 keV (0.061 %)
Pitch factor	g_{O}	1.22 (8.6 %)
Axial velocity component	β_z	0.327 (5.0 %)
Perp. velocity component	$\beta_\perp$	0.398 (3.4 %)
Beam radius	r_{b}	10.24 mm (0.87 %)
Magnetic field at emitter	$B_{\mathrm{E}(,z)}$	0.1909 T
Magnetic compression (field)	C_{B}	50.2
Magnetic compression (beam)	$C_{r_{\mathrm{b}}}$	40.3

2.4 Further Key Component Considerations

Launchers for coaxial-cavity gyrotrons have been designed for one-beam [PBD+99, RPK+10] and two-beam [FKM+94, PBD+98, Ber11] output. Independent of the launcher type, their fundamental design depends on the caustic radius of the operating mode and scales roughly with the cavity diameter. Both cavity and main mode presented in this work have properties typical for high-power high-frequency gyrotrons; therefore, problems with the basic launcher design are not expected. The operating wavelength of 1.26 mm is substantially shorter than the 1.76 mm of the 170 GHz ITER gyrotrons, which could lead to difficulties in manufacturing of in-launcher corrugations. However, during the design of a conventional-cavity launcher for 236 GHz [KAF+16b], no potential problems were observed. As the corrugation size in this case is up to 450 μm and as the machining tolerance is well below 20 μm, no manufacturing issues are expected.

In the magnet specifications, criterion #5 (see Tab. 2.6 and Fig. 2.7) originates from the requirement that the beam should not decompress so quickly that electrons hit the upper part of the launcher. However, according to present understanding it would be difficult for magnet manufacturers to actually miss that requirement.

Due to the comparable launcher design, the design of the **mirror system** is not expected to differ much from existing designs. But, since a large gyrotron magnet is required, it could happen that the mirror box has to be placed partially within the magnet. Possible solutions would be to increase the end diameter of the launcher, leading to a higher k_z of the wave and thus to a longer beam output system, or to require the bore-hole of the magnet to have a larger diameter at the top than it has in the center. Detailed launcher and mirror box considerations on the one hand, and detailed magnet design studies on the other hand would be necessary to obtain specific values here, which might result in a compromise between a convenient mirror box design and a simple magnet design.

The **output window** for the gyrotron design in this work would not have another thickness or would have to withstand more radiation power than the windows currently considered for the ITER 2 MW coaxial gyrotron [RPK$^+$10]. Advantages of Brewster-angle windows are independent of the incident frequency. Therefore, a window for the presented 238 GHz design does not need special investigation.

A standard single-stage depressed **collector** could be used for the gyrotron presented in this work, possibly with stronger coils and/or thicker ferromagnetic shielding to compensate the high residual magnetic field from the gyrotron main magnet. But since the interaction efficiency does not exceed 33 %, an SDC is not sufficient to increase the total gyrotron efficiency to above 50 %.

Efficient multi-stage depressed collectors to increase the total efficiency to above 60 %, such as proposed in [WPI$^+$16a, WPI$^+$16b, PWI$^+$16], rely on sophisticated static electric and magnetic field configurations and are therefore less robust against residual fields than SDCs [Wu16]. This issue could be mitigated if the superconducting magnet for a high-frequency gyrotron is already known and can be considered during the MSDC design. Still, such collectors would have a significant size (length ca. 2 m) and weight (ca. 1 t) and could be a reason to restrict the magnet bore-hole size and thus the gyrotron output power in order to obtain higher total efficiency.

2.5 System Evaluation

The evaluation of two features of the gyrotron design are of particular interest: First, whether the system indeed supports operation at distinct frequencies (as promised in Section 2.1.3) and second, how the output power could possibly be increased.

2.5.1 Multi-Frequency Behavior

As introduced in Section 1.2.3 one of the gyrotron design criteria was suitability for multi-frequency operation, resulting in the mode series $TE_{35,21}$ – $TE_{42,25}$ – $TE_{49,29}$ (see Section 2.1.3). With the finished physical design of MIG, magnet, and cavity, one can now verify in self-consistent beam optics simulations in conjunction with time-dependent multimode interaction simulations with realistic beam parameters whether startup and stable operation at the various frequencies can be achieved and whether obstacles occur which could not be considered at the mode-selection stage.

Tables 2.8-2.9 reproduces the operating points and simulation results for all three frequencies under consideration [FAP+15, FPA+15a, FPA+15b]; see also Tables 2.1 to 2.7. Despite the significantly different operating frequencies, the overall magnetic field profile and consequently the optimum electron beam radius in the cavity is almost constant; minor changes to the profile, including magnetic compression, can be made by optimizing the currents of the compensation and gun coils. On the other hand, the various magnetic field strengths in the emitter region require substantially different modulation anode voltages (percentages of the $TE_{49,29}$ values are given in Tables 2.8-2.9) in order to keep the pitch factor at reasonable values, demonstrating that a diode MIG will probably not be suitable for a multi-frequency gyrotron. In the $TE_{35,21}$ mode scenario, the anode-anode voltage is even significantly larger than the modulation anode; both voltages have essentially switched compared to the $TE_{49,29}$ case. Cavity pitch factor, beam radius, electron energy, and beam current have been optimized for all three operating points under the assumption that accelerating voltage and current density corresponding to the $TE_{49,29}$ operating point are at the technical limits of the gyrotron's HVDC power supply.

The MIG simulations returned velocity spreads from 3.4 % to 6.8 %. Included here is a typical emitter surface roughness of 2 µm, but no other sources for velocity spread such as azimuthally or radially inhomogeneous emission caused by material impurities, varying work function, or varying emitter temperature. In order to account for this, the input velocity spread for interaction simulations was increased to 6 % and 8 %, respectively.

The voltage depression on the beam is only around 2 kV for all three operating points. The change in relative beam thickness is only due to the significantly varying wavelength; thus, a tube optimized for 170 GHz would allow larger output power since it could have a broader emitter which can deliver more current.

The beam shape for all three frequencies is shown in Fig. 2.13. Despite huge leaps in magnetic fields and voltages between the scenarios, the beam shapes remain similar. However, with the present geometry and even without the maximum electric field restriction of 7 kV/mm it was not possible to obtain a stable beam for the next frequency step, $TE_{56,33}$ (271.3 GHz), due to the high magnetic field strengths near the cathode nose (see Fig. 2.11 at (-0.375;0.055)). This, together with the required emitter thickness (at most 3.8 mm) and magnetic field (11 T) led to the conclusion that the $TE_{56,33}$ mode operation regime is beyond technical feasibility. On the other hand, operation of mode $TE_{28,17}$ (136.2 GHz) would be neither restricted by the magnet nor by the interaction itself, but the spread of the electron velocities is far above 10 % without modification of the MIG design.

The most serious issue concerning coaxial-cavity multi-frequency gyrotrons seems to be ohmic loading on the insert wall at the lower frequencies. For lower frequencies one would expect lower losses (see Annex B.7), which is indeed the case for the losses on the cavity wall. In contrast, the loading on the insert is significantly higher for the lower-order modes due to their broader first maximum and due to the fact that the field strength decreases exponentially towards the insert, see Table 2.10. This means that either the output power (the beam current) has to be reduced substantially or that the insert radius has to be modified.

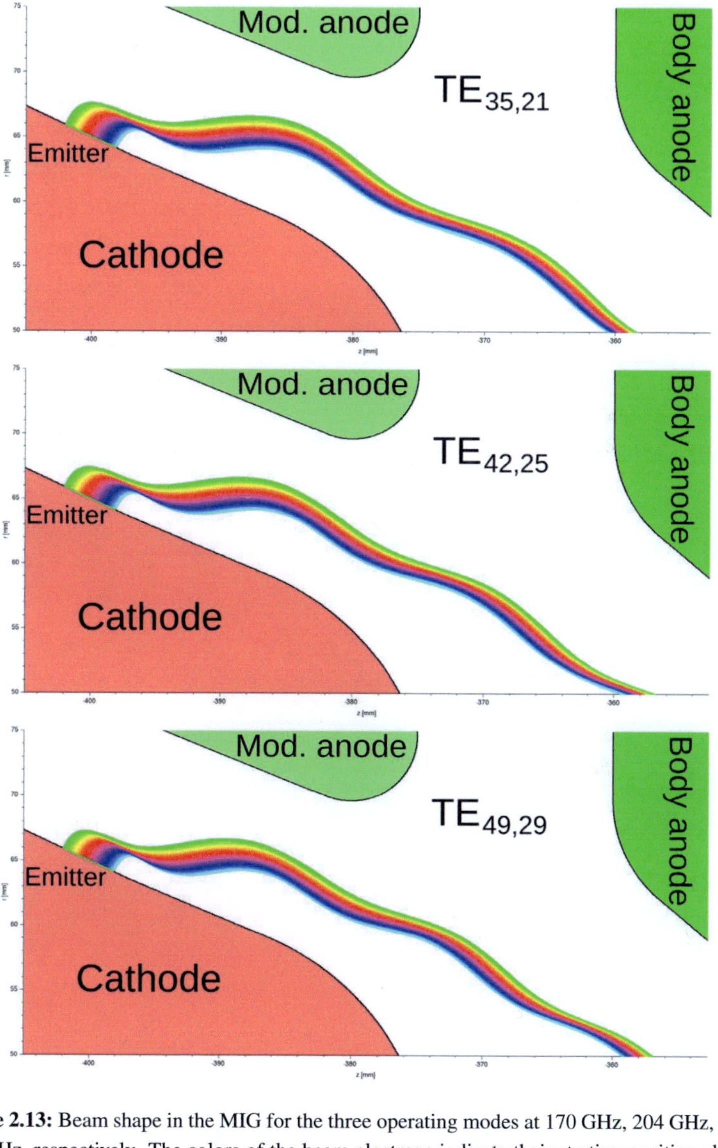

Figure 2.13: Beam shape in the MIG for the three operating modes at 170 GHz, 204 GHz, and 238 GHz, respectively. The colors of the beam electrons indicate their starting position along the emitter.

If reducing the output power is not an option, one comparably realistic solution to this issue could be to use an axially movable insert with varying radius, similar to [IDT08], since the insert has to be tapered anyway as explained in Annex B.4, see Fig. 2.14. With a typical taper angle of 1° the necessary shift would be 5 mm to 6 mm. As indicated in the Figure, one could then also apply various corrugations on the various parts of the insert, each optimized for the respective mode and wavelength: 0.35 mm depth for $TE_{42,25}$, and 0.48 mm depth for $TE_{35,21}$. This method would, however, most possibly increase insert tilt and misalignment, which is not favourable, as will be discussed in Chapter 3.

See Fig. 1.7 for the overall design of a gyrotron with MIG, magnet and cavity as described in this Section.

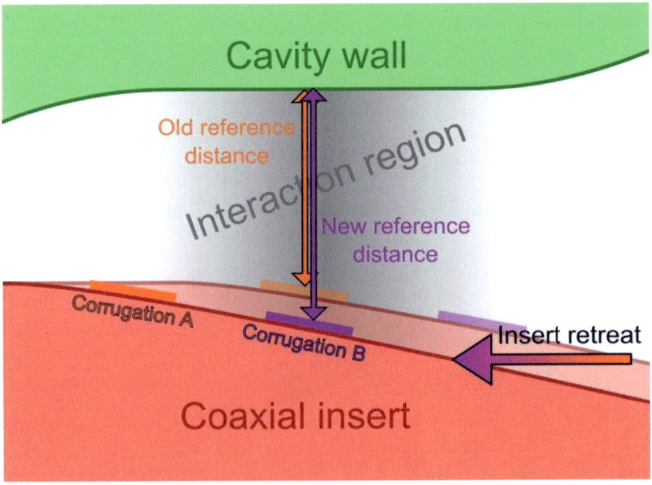

Figure 2.14: Principle of a movable tapered insert: The distance between cavity wall and insert at the reference point (roughly in the middle of the interaction region) can be changed, leading to different mode patterns and to a changed ohmic loading on the coaxial insert. If different corrugations are applied in different places of the insert, the effective corrugation (again, near the middle of the interaction region) can be altered as well.

Table 2.8: Operating parameters of the three designated operating modes, 1.

Relative window thickness	$2D_\mathrm{W}/\lambda$	**5**	**6**	**7**
Operating mode		$\mathrm{TE}_{35,21}$	$\mathrm{TE}_{42,25}$	$\mathrm{TE}_{49,29}$
Mode eigenvalue	χ	113.1	135.6	158.1
Frequency (GHz)	f_D	170.0	203.8	237.5
Magnetic field at emitter (T)	B_E	0.137 (71.7 %)	0.165 (86.4 %)	0.191
Accelerating voltage (kV)	U_EA	85.6 (97.9 %)	87.9 (100.6 %)	87.4
Modulation anode voltage (kV)	U_EM	31.9 (63.9 %)	41.3 (82.8 %)	49.9
Anode-anode voltage (kV)	U_MA	53.7 (143.2 %)	46.6 (124.3 %)	37.5
Beam current (A)	I_b	69.4	70.0	69.3
Velocity spread in cavity by MIG	ζ_{β_z}	6.8 %	4.3 %	3.4 %
Magnetic field in the cavity (T)	B_O	6.82	8.22	9.58
Magnetic compression (field)	C_B	49.8	49.8	50.2
Electrons' kinetic energy (keV)	W_e	83.4	86.0	85.6

Table 2.9: Operating parameters of the three designated operating modes, 2.

Operating mode		$\mathrm{TE}_{35,21}$	$\mathrm{TE}_{42,25}$	$\mathrm{TE}_{49,29}$
Velocity spread (interaction sim. input)	$\zeta_{\beta_z,\mathrm{in}}$	8 %	6 %	6 %
Guiding center radius in cavity (mm)	r_b	10.28	10.27	10.24
Magnetic compression (beam)	C_{r_b}	40.0	40.1	40.3
Pitch factor in cavity center	g_O	1.27	1.25	1.22
Relative beam thickness	$\varsigma_{r_\mathrm{b}}/\lambda$	6.2	5.1	4.4
Output power (MW)	P_out	1.8	1.9	1.9
Electronic efficiency	η_elec	31 %	32 %	33 %
Main coil current (A)	I_MC	101.8 (71.2 %)	122.7 (85.9 %)	142.9
Compensation coil 1 current (A)	I_CC1	-104.6 (70.9 %)	-126.2 (85.6 %)	-147.5
Compensation coil 2 current (A)	I_CC2	-98.6 (71.5 %)	-118.7 (86.1 %)	-137.9
Gun coil current (A)	I_GC	-72.5 (70.5 %)	-87.6 (85.2 %)	-102.8

Table 2.10: Peak ohmic loading on the insert for the three operating points.

Operating mode		$TE_{35,21}$	$TE_{42,25}$	$TE_{49,29}$
Peak ohmic wall loading (kW/cm²)	$\hat{w}_{\Omega,O}$	1.3	1.7	2.0
Total ohmic loss on wall (kW)	$P_{\Omega,O}$			47.6
Total ohmic loss on insert (kW)	$P_{\Omega,I}$			1.0
Corrugation depth (mm)	$D_{r,I}$	0.3	0.3	0.3
Corrugation width (mm)	$D_{\varphi,I}$	0.3	0.3	0.3
Constant radius, varying loading				
Insert radius in central cavity (mm)	R_I	8.6	8.6	8.6
Peak ohmic loading (kW/cm²)	$\hat{w}_{\Omega,I}$	0.5	0.3	0.2
Varying radius, constant loading				
Insert radius in central cavity (mm)	R_I	8.4	8.5	8.6
Peak ohmic loading (kW/cm²)	$\hat{w}_{\Omega,I}$	0.2	0.2	0.2

2.5.2 Output Power at Higher Beam Current and Higher Loading

The presented design can reach an output power of close to 2 MW. However, it is a general observation that MW-class gyrotrons do not reach their full design power under realistic conditions. Therefore, if the proposed gyrotron was built as presented, one should not expect more than 1.8 MW at 238 GHz.

The critical limitations on the output power are beam current density and the capabilities of the cavity, insert, and collector cooling systems. As mentioned in Section 2.1.1, conservative values for current density and ohmic loading have been assumed. A current density of $4.5\ \mathrm{A/cm^2}$ will not impede the emitter lifetime significantly. Simple estimations predict that an electric conductivity of around $2 \cdot 10^7\ \mathrm{S/m}$ can be achieved in the cavity wall (copper at $250\ ^\circ\mathrm{C}$ and 240 GHz); with the possible output power scaling with $\sqrt{\sigma}$, i.e. $20\ \%$ higher (see Annex B.7). Thus, a moderate increase of the beam current by $11\ \%$ could lead to gyrotron operation at 2 MW.

Deliberately pursuing a $\mathrm{TE}_{49,29}$ mode design with significantly more than 2 MW continuous-wave output power based on the presented design appears to be problematic. First, the necessary beam energy must result from a higher accelerating voltage, implying a new – probably larger – MIG design, and/or a higher beam current, which in turn requires increased emitter current density and/or a larger emitter radius. As a result, one would have to push current density and magnet bore-hole diameter beyond even the optimistic values stated above and in Section 2.3.2, respectively. Second, one would need an additional mirror system and diamond window (Brewster-angle window for a step-tunable gyrotron) without even remotely using their full capability of additionally 2 MW. Third, the multi-stage collector, already assumed to be of critical size (see Section 2.4), would have to be scaled accordingly. For these reasons, a realistic 240 GHz, $>$ 2 MW gyrotron design needs in-depth technical considerations (and solutions), alongside a reliable experimental basis e.g. from CW gyrotrons above 1 MW and 200 GHz and with large magnetic compression.

2.6 Recommendations for Future Gyrotron Design Processes

During the design of the gyrotron presented in this work a few measures have been identified which can make the design process more efficient:

- It is of utmost importance to agree on design goals and on technical boundary conditions from the very beginning on. An existing physical gyrotron design can of course be adapted to, for example, a higher emitter current density or to a higher efficiency, but obtaining a good initial design takes much less effort than optimization of all components, which is necessary for most minor changes of the boundary conditions.

- For otherwise similar designs, the expected output power is proportional to the allowed ohmic losses in the cavity and to the emitter current density. Thus, two designs can be comparable only if the underlying assumptions for both are identical, which, as stated above, should be defined at an early stage.

- Output power and efficiency depend significantly on the choice of beam parameters such as beam thickness, electron energy and velocity spreads, and misalignment. Optimization of the gyrotron design is usually done using ideal beam parameters (no thickness, spread or misalignment), but its outcome might differ from an optimization using realistic parameters. Therefore, estimated values for realistic beams should be included in the cavity design from the beginning.

- For MIG and cavity optimization, automatic procedures should be developed and used more frequently, e.g. based on an evolutionary algorithm or on simulated annealing, as has been done in the past [RBI+09]. These procedures should at best be independent of the type of MIG (e.g. diode/triode, inverse) and cavity (e.g. coaxial, tapered).

- For coaxial-cavity gyrotrons with corrugations the acutal effect of the corrugations should be determined using different corrugation models, e.g. SIM, SHM, and SIE. If only one model is used, it should be benchmarked to an eigenmode solver.

3 Design Considerations for Inner Conductor and Cavity

3.1 Coaxial Cavity with Broken Axisymmetry

This Chapter is dedicated to an in-depth evaluation of the design decribed in the previous Chapter. As the topic of the present work is a coaxial-cavity gyrotron design for unprecedentedly high frequencies, the design evaluation will primarily focus on issues which are concerned with both the high frequency (or, equivalently, with the small wavelength) and the coaxial insert. Two specific questions of more general interest, which also arise during the design of conventional gyrotrons and/or gyrotrons at lower frequencies, will be adressed in Sections 3.4 and 3.5. Other topics, which are not included here, would encompass detailed studies of emitter properties and behavior, emitter heating, optimization of beam tunnels, thermal expansion of components in detail, BCI and ACI, launcher and collector design, cooling, mechanical stability, questions concerning operation and maintenance, and alignment issues not concerned with the coaxial insert, only to mention a few. As expounded in the previous Chapter, the presented design – apart from some aspects of the magnet – was kept at conservative values and it does not contain any unrealistic assumptions that would require additional in-depth studies. From these considerations, two key questions remain, both of which cannot be answered if axisymmetry of the cavity is assumed.

The first question is which size and shape the longitudinal corrugations of the insert should have. In reality, milled corrugations have parallel walls and more or less rounded edges. However, for practical reasons, only wedge-shaped corrugations are considered in this work, with the walls pointing exactly in radial direction with respect to the insert center, and with concentric circle segments as top and bottom surfaces. These top and bottom parts are always assumed to be of equal arc length. Thus, the corrugations can be described uniquely by only two numbers: corrugation depth $D_{r,\mathrm{I}}$ and corrugation number M_{I}. As has been investigated a few years ago, the actual shape of the corrugations does not influence the mode eigenvalues and quality factors, and usual corrugations can be described by wedge-shaped corrugations with a certain effective depth [ZM11].

In the presented interaction calculations, the Surface Impedance Model (SIM; see [Ker96] and Annex B.3) has been used exclusively. In this model the insert surface is regarded locally as a plane with rectangular grooves, the number of which is large enough such that the groove width is sufficiently smaller than the wavelength. The only free parameter describing the corrugations is then their (effective) depth $D_{r,\mathrm{I}}$. More accurate modelling could be done by more advanced models such as Space Harmonic Method (SHM) and/or Singular Integral Equation (SIE) [ZGS04]. Alternatively, there are more elaborate variants of the standard SIM [IALT11].

In the limit of small corrugation width, wedge-shaped corrugations, rectangular corrugations, and corrugations of realistic shape are equivalent. The only remaining variable to be determined is then M_{I}. For a technical cavity design, the number of corrugations would have to be determined by practical considerations, i.e. how thin those corrugations can be made. The corrugation width will be considered small enough (and M_{I} considered large enough) if SIM and direct field calculations are in sufficient agreement.

The second question is which influence tilt and shift of the insert have on gyrotron interaction, considering the small operating wavelength of only 1.26 mm, and how to cope with such misalignment. In fact, only the shift needs to be considered, since, with an emitter-cavity distance of e.g. 412 mm, a tilt of $0.14°$ of the insert at its base is equivalent to at least 1 mm displacement in the cavity, while the tilt itself is negligible. This can therefore be perfectly described by a shift only, and two-dimensional simulations are widely sufficient. Conversely, an intrinsic tilt of the insert can be compensated by a proper shift, using the same argument as in [DAFJ14], table III for a tilted electron beam. Concerning the position of the electron beam within the coaxial cavity, it is assumed that gyrotron and magnet can be shifted towards each other by at least by 1 mm such that the beam can be aligned to the insert or to the cavity wall (or neither). In the following, the displacement between insert and wall is expressed as D_{IO}, between insert and beam as D_{Ib} and between beam and wall as D_{bO}.

3.2 Required Number of Corrugations

The final number of corrugations on the coaxial insert, M_{I}, is a tradeoff between a number for which the SIM and an actual calculation of the mode patterns and eigenvalues correspond to each other (ideally: ∞) and a corrugation number which is most easy to be realized (ideally: 0). In the mechanical workshop of IHM corrugations having a width $D_{\varphi,\mathrm{I}}$ of 0.3 mm can be manufactured without problems, while in the central workshop at KIT widths of down to 0.2 mm are considered possible [Kob15]. With an insert radius of 8.6 mm, $D_{\varphi,\mathrm{I}} = 0.3\,\mathrm{mm}$ would correspond to $M_{\mathrm{I}} = 90$ corrugations around the circumference. In this Section, the optimum number of corrugations for the given geometry will be deduced, based on very general considerations. In order to distinguish between corresponding modes in geometries with different M_{I}, the notation $\mathrm{TE}_{mp}\{M_{\mathrm{I}}\}$ will be used.

Calculations have been done using the two-dimensional finite element (FEM) eigenfrequency solver of COMSOL. The solver returns two orthogonal stationary solutions with the same frequency for asymmetric ($|m| > 0$) modes, from which the rotating modes could be constructed by time-dependent linear superposition, and one solution for symmetric ($m = 0$) modes. Perfectly conducting metallic surfaces were assumed. Unless stated otherwise, plots show the magnitude of the local transverse electric field. In order to obtain meaningful results for corrugation numbers up to 360 (regarded as “large enough”), the mesh size has been set to 20 micrometers within the corrugations and to less than 70 micrometers in more homogeneous regions, with a small growth factor inbetween. The solver can return eigensolutions of any 2D shape and does therefore not rely on eigenmodes of pre-defined geometries. This has the advantage that its solutions can readily be benchmarked with analytic calculations, but also has the disadvantage that there is no possibility to directly calculate the (hollow-cavity) mode content of solutions. However, one has to keep in mind that the output results *are* eigensolutions of the problem, each with one clearly defined eigenfrequency. Only in the case where the coaxial insert disappears in the direction of propagation of the mode, as it happens at a certain z-position in the uptaper/launcher region, the wave would collapse into a number of hollow-cavity modes, each with its characteristic cutoff frequency, transversal and longitudinal structure, in particular $\mathrm{TE}_{mpl} \mapsto \mathrm{TE}_{mp'l'}$. Figure 3.1 shows the used mesh during the calculations and some details of one selected solution. The mode on the right is $\mathrm{TE}_{45,30}\{100\}$, which has a rather small caustic radius and therefore its field maximum is close to the insert. Mind that even though the aperiodic coupling between mode and insert is strong, the mode pattern above 0.3 mm (one corrugation depth) from the insert is not visibly influenced by the corrugations.

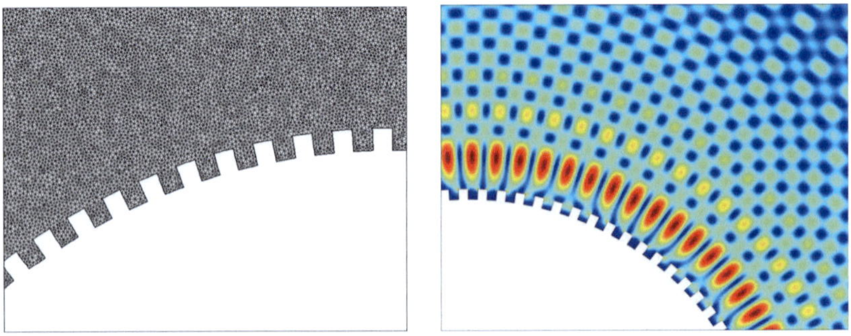

Figure 3.1: COMSOL triangular mesh (left) and simulation results (right) of eigenmode calculations with 100 corrugations on the insert. See text for details.

For small numbers of corrugations, primes were used to avoid resonant effects that might have occured if corrugation number and azimuthal mode index were not prime relative to each other. Since there were no differences in results between simulations with 100 corrugations and with 101 corrugations, more convenient numbers for $M_{\mathrm{I}} > 100$ were used. In the subsequent misalignment studies, the corrugations were aligned antisymmetric with the direction of insert-wall misalignment (i.e., the misalignment direction points in the direction of a radial wall of one corrugation), see Fig. 3.2, to represent an average/random twist of the insert. The opposite cases – symmetric alignment with a top or a bottom wall in misalignment direction – led to eigenmodes which have larger or smaller eigenfrequencies; however, this difference exceeds a noticeable relative frequency shift of 10^{-7} only for much smaller wavelengths, much larger corrugations, and smaller relative caustic radii than in the scenarios discussed here.

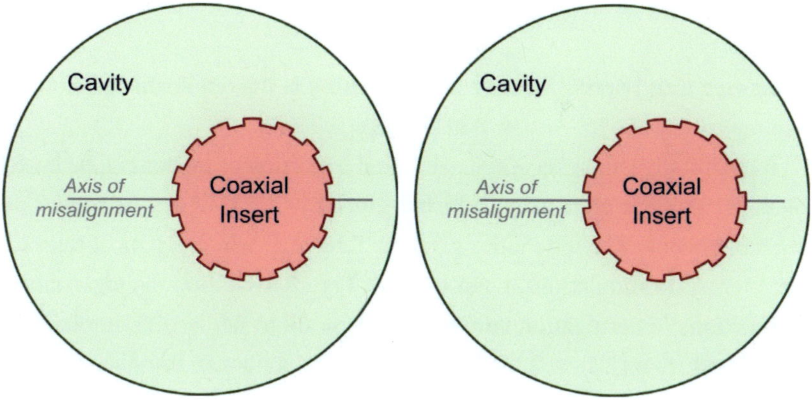

Figure 3.2: Schematic representation of the insert twist versus direction of misalignment as used in the misalignment simulations (left) and a possible twist not considered (right).

If the number of corrugations is too small, resonance effects are expected to occur. Those effects can be understood as coupling between high-order and low-order modes (as would be intended in rippled-wall corrugations

[Ker96]), as interferences caused by Bragg reflections, or as interaction between two periodic structures, the azimuthal mode maxima, and the grooves. For an otherwise constant coaxial-cavity geometry (including a constant caustic radius close to the insert radius), it follows from basic geometrical considerations that wavelengths and azimuthal indices of modes have approximately linear dependence, and therefore it is intrinsically impossible to distinguish between effects caused by longitudinal dimensions of the same order ($\lambda \sim D_{\varphi,\mathrm{I}}$) and by angular dimensions of the same order ($m \sim M_\mathrm{I}$). Thus, one can conclude that the various interpretations of the resonance effects are equivalent.

Figure 3.3 illustrates the effect of different corrugation numbers: while mode $\mathrm{TE}_{46,30}\{89\}$ has visible contributions of field patterns with rotational symmetries 3 and 9, $\mathrm{TE}_{46,30}\{100\}$ appears fully symmetric. It has been found that for most modes, the relation for constructive higher resonances

$$|2m - M_\mathrm{I}| = q \tag{3.1}$$

holds (see also [Ker96], section 2.4.), where q is the dominant lower-order symmetry, e.g. $2 \cdot 46 - 89 = 3$ in the depicted example.

A quantitative analysis of the azimuthal distortion of modes can be based on their change in eigenvalue, which is given by COMSOL as the respective eigenfrequency. The ten modes given in Table 1.1, which are most relevant for interaction simulations, were studied. Figure 3.4 shows the eigenvalues as functions of corrugation number for $M_\mathrm{I} = 59$ to $M_\mathrm{I} = 225$, normalized to their values for $M_\mathrm{I} = 225$. The following behavior can be observed:

- Modes with smaller caustic radius, i.e. with higher field near the insert, are generally more affected by small corrugation numbers than modes with larger caustic radius, as they should be by all changes to the insert.

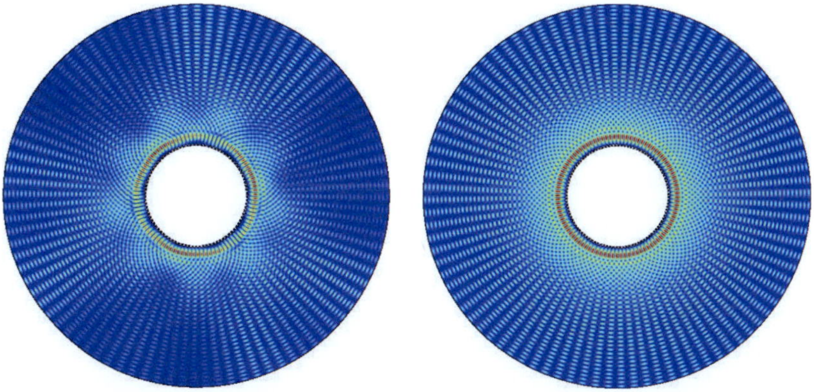

Figure 3.3: Mode $\mathrm{TE}_{46,30}$ in a coaxial cavity with 89 corrugations (left) respectively 100 corrugations (right) on the insert.

- For large M_I, eigenvalues increase with increasing M_I. This can be understood if the insert surface is considered as an intermediate state between a perfect metallic surface ($M_\mathrm{I} \to 0$) and a surface with perfect impedance ($M_\mathrm{I} \to \infty$) of the same radius: the larger the contribution of the impedance is, the more can the field penetrate through the surface and the larger is the effective cavity cross-section covered by the mode, corresponding to a larger eigenvalue. However, as one can see from the numbers, this effect is below 0.1 %.

- A linear increase in M_I will lead to a less-than-linear increase of the mode eigenvalue ($\mathrm{d}^2\chi_{mp}/\mathrm{d}{M_\mathrm{I}}^2 < 0$), which is due to the infinitesimal nature of the impedance surface. Mind that the normalization in Fig. 3.4 is arbitrarily chosen to be at $M_\mathrm{I} = 225$.

- The changes in eigenvalue are homogeneous in corrugation number M_I and relative caustic radius C_R only for $M_\mathrm{I} \gtrsim 100$. Smaller corrugation numbers result in less predictable behavior, probably due to occurring resonances with the respective azimuthal index as soon as wavelength and groove width are of the same order.

For large corrugation numbers M_{I}, the COMSOL solutions should approach the SIM predictions as the latter assumes $M_{\mathrm{I}} \to \infty$. In Table 3.1, eigenfrequencies of the ten modes for $M_{\mathrm{I}} = 360$ are shown. Exact geometric values, expressed with the same precision as the stated results, are $R_{\mathrm{O}} = 31.78000$ mm, $R_{\mathrm{I}} = 8.589292$ mm, $D_{r,\mathrm{I}} = 0.3009566$ mm. For $M_{\mathrm{I}} = 360$, the exact frequencies f_{360} are still below, but already very close to their SIM values f_{SIM}. The features observed and explained above, that eigenvalues of modes with smaller caustic differ more from their SIM values and that the SIM values are always higher, are preserved. Therefore, the presented results, SIM as well as COMSOL, can be considered verified. Mode $\mathrm{TE}_{45,30}$ will be the main focus of the following studies, since it has the smallest caustic radius $C_{\mathrm{K}} \equiv \frac{m}{\chi_{mp}}$ of the ten modes chosen, leading to highest sensitivity to the number of corrugations and misalignment.

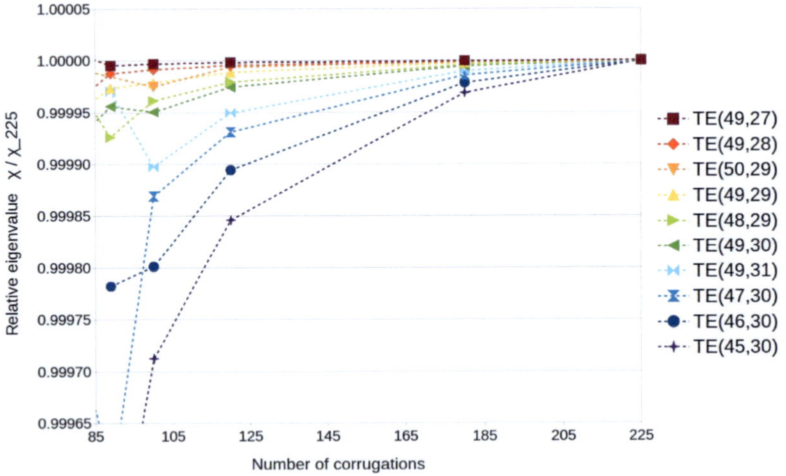

Figure 3.4: Eigenvalues vs. corrugation number for ten modes using COMSOL. Modes are sorted according to their relative caustic radius, with $\mathrm{TE}_{45,30}$ having the smallest radius.

Table 3.1: Comparison between eigenfrequencies determined with the SIM and those determined with the COMSOL eigenfrequency solver.

m	p	$\frac{m}{\chi_{mp}}$	f_{SIM} (GHz)	f_{360} (GHz)	f_{360}/f_{SIM}
49	27	0.324	227.3593	227.3593	0.9999999
49	28	0.317	232.3405	232.3403	0.9999991
50	29	0.314	239.2895	227.2891	0.9999985
49	29	0.310	237.3132	227.3125	0.9999970
48	29	0.306	235.3375	235.3360	0.9999936
49	30	0.304	242.2821	242.2803	0.9999924
49	31	0.298	247.2541	247.2505	0.9999853
47	30	0.296	238.3464	238.3413	0.9999786
46	30	0.292	236.3972	236.3895	0.9999673
45	30	0.288	234.4694	234.4458	0.9998995

A more detailed quantitative analysis of very high-order modes in waveguides with corrugated insert has been developed, based on spacial Fourier harmonics. In the following, the transverse electric field of the modes was understood as $E_\perp = E_r + \mathrm{i}E_\varphi$, where both components are functions of the radial coordinate r and of the azimuthal coordinate φ. The radial coordinate r has been fixed to $r_\mathrm{b} \equiv 10.24$ mm, therefore, the 2π-periodic, complex function $E_\perp(\varphi)$ remained. (Note that this function differs from $E_\perp \equiv E_x + \mathrm{i}E_y$ in Annex A by a factor of $e^{\mathrm{i}\varphi}$ in order to account for "unwinding" the azimuthal cordinate.) This function can now be written in the form

$$E_\perp(\varphi) = \sum_{q=0}^{\infty} E_q e^{\mathrm{i}q\varphi} \tag{3.2}$$

where E_q are complex Fourier coefficients. The spectrum of $|E_q|^2$ over q gives the relative power of q-fold rotational symmetric contributions. It is obvious that an axisymmetric field pattern would give $E_\perp(\varphi) = E_0$ and a constant field would give a one-fold symmetric pattern (plus higher harmonics), which is consistent with the intuitive understanding of the contributions of the E_q. Along the azimuthal coordinate a large prime number of discretization steps, namely $n_\varphi = 59359$, was chosen to avoid common divisors or multiples of the step number and the azimuthal index, which might lead to numerical errors. Due to $2 \cdot 45 - 89 = 1$ (Eq. 3.1), it is clear that $\mathrm{TE}_{45,30}\{89\}$ would not have been a good choice for illustration in Fig. 3.3, since it simply adds a linearly increasing field over the cavity cross-section.

For illustration, Fig. 3.5 shows three power spectra up to about $q = 80$. As one can see (and expect), $\mathrm{TE}_{45,30}\{100\}$ has one large amplitude for $q = m = 45$; the other contributions, e.g. the resonance at $q = 55 = M_\mathrm{I} - m$, are significantly lower. Mode $\mathrm{TE}_{46,30}\{89\}$, which was shown on the left in Fig. 3.3, has large contributions of $q = m = 46$, $q = 43 = 89 - 46$, and $q = 37$; and smaller ones at $q = 20$ (which might be related to the number of azimuthal discretizations n_φ), $q = 52$, and $q = 54$. The visible pattern in Fig. 3.3 could result from the differences $\Delta q = 46 - 43 = 3$ and $\Delta q = 46 - 37 = 9$. Mode $\mathrm{TE}_{46,30}\{100\}$ has only the contributions $q = m = 46$ and, at smaller power, $q = 20$ and $q = 100 - 46 = 54$ left.

Figure 3.6 shows power spectra for mode $\mathrm{TE}_{45,30}\{M_\mathrm{I}\}$ with selected M_I. For $M_\mathrm{I} < 100$, the spectra have strong components in addition to $q = 45$, in particular $q = 14 = 59 - 45$ for $\mathrm{TE}_{45,30}\{59\}$ and $q = 34 = 79 - 45$ for $\mathrm{TE}_{45,30}\{79\}$. For $M_\mathrm{I} > 100$, only $q = 25$ remains strong for all M_I, but not significant. All other spectral components are not far above the numerical noise level, as obtained by spectral analysis of mode $\mathrm{TE}_{45,30}\{0\}$, i.e. with a smooth insert. Hence, one can conclude that $M_\mathrm{I} = 100$ is sufficient for the geometry considered in Chapter 2, in the sense that the eigenvalues and transversal profiles of the main mode and of its most severe competi-

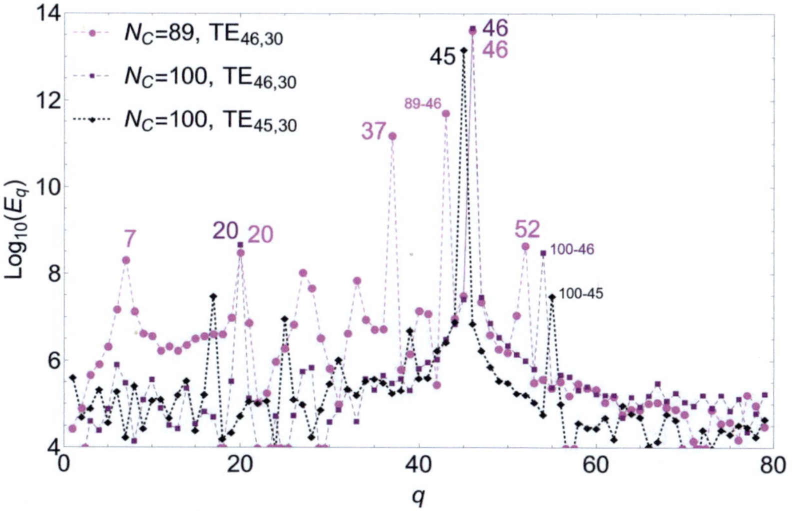

Figure 3.5: Power spectra for modes $\mathrm{TE}_{45,30}\{100\}$, $\mathrm{TE}_{46,30}\{89\}$, and $\mathrm{TE}_{46,30}\{100\}$.

tors are sufficiently well predicted by the SIM. The much stronger criterion $M_\mathrm{I} > |m| + \chi_{mp}$ ([IALT11], Eq. 14), which would demand more than 200 corrugations, is thus not considered mandatory.

With $R_\mathrm{I} \approx 8.6$ mm and $D_{\varphi,\mathrm{I,min}} \approx 0.2$ mm, one could safely manufacture $M_\mathrm{I,max} \approx 135$ corrugations with the required dimensions. Therefore, the corrugation number analysis in this Section supports the design described in the previous Chapter.

3.3 In-Depth Analysis of Insert Misalignment

3.3.1 Overview

Being the desired set-up, perfectly aligned geometries, i.e. geometries where emitter, insert (if present), electron beam, and cavity wall are exactly concentric versus each other, are usually considered in hollow- and coaxial-cavity gyrotron studies. In reality, however, concentricity can only

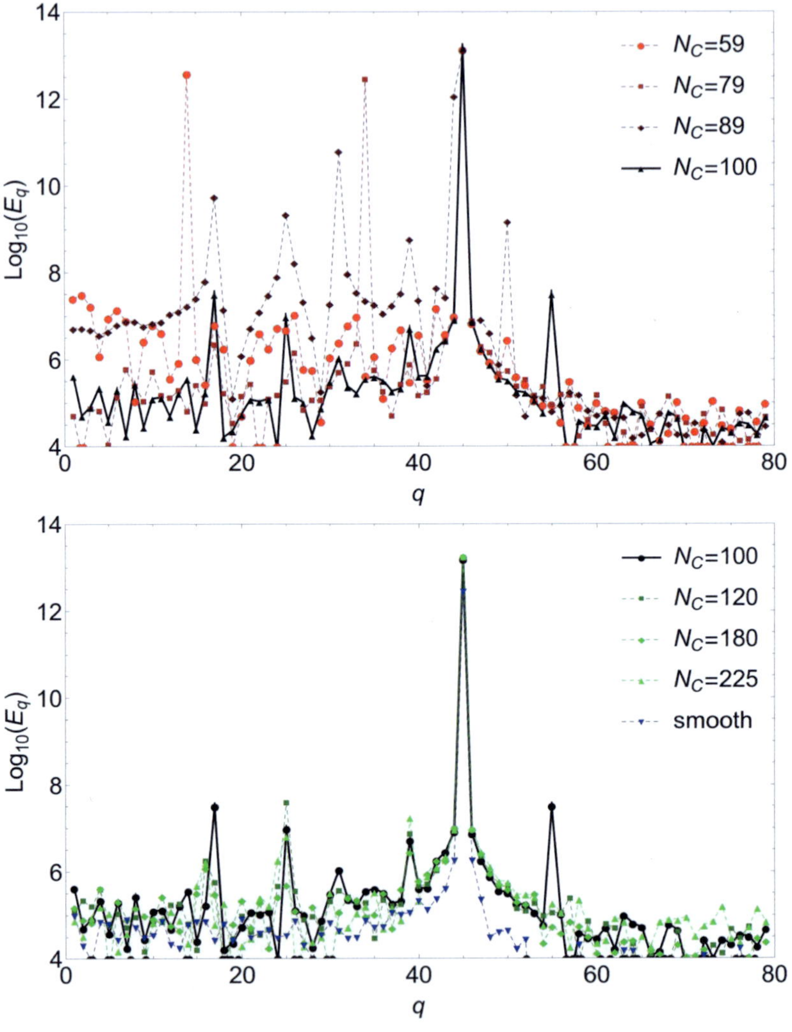

Figure 3.6: Power spectra (real parts of the amplitudes) for $\mathrm{TE}_{45,30}\{M_\mathrm{I}\}$ with $M_\mathrm{I} \leq 100$ (upper) and $M_\mathrm{I} \geq 100$ (lower). See text for discussion.

be achieved to a certain extent. Even if one assumes perfectly cylindrical components (no elliptical deformations, no unintended bends), misalignments – lateral displacements (shifts) or angular displacements (tilts) – between the axes of the following components will inevitably occur:

- Magnet,
- Magnetic field axis (equivalently: electron beam axis) at the cavity,
- Magnetic field axis at the emitter,
- Gyrotron body (as a whole),
- Cathode with emitter,
- Cavity wall, and
- Coaxial insert.

Gyrotron body axis, emitter axis, cavity axis and – commonly – insert axis are fixed against each other, as are the mechanical magnet axis and the magnetic field axis at the cavity. However, gyrotron body and magnet can usually be moved versus each other such that the magnetic field axis at the emitter coincides with the emitter axis. Tilting the gyrotron or using dipole magnets of sufficient strength, the magnetic field axis in the cavity region could now be aligned with either the cavity wall axis or with the insert axis (or with neither of both). See Fig. 3.7 for an illustration: Only if the insert can be aligned and if dipole coils are used, all field axes (red, yellow) can coincide with mechanical axes (green, blue). Assuming that cavity wall and insert cannot be aligned, the following questions are to be answered:

1. Should the electron beam be aligned with the wall, with the insert or neither?
2. How can the effects of this misalignment on electron beam and RF field be simulated?

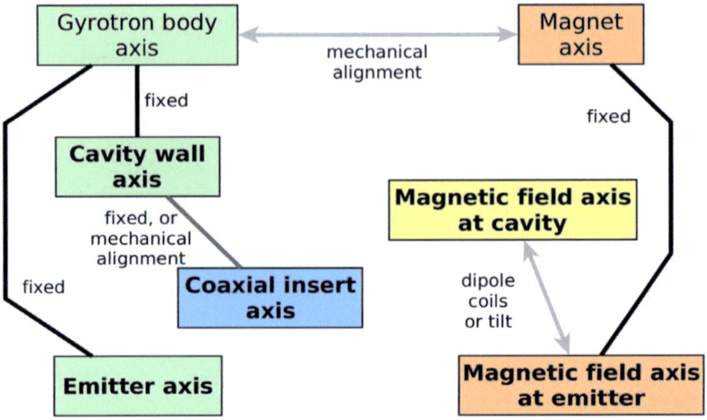

Figure 3.7: Possibilities to align gyrotron components. *Green:* Mechanical gyrotron structure. *Blue:* Coaxial insert. *Red:* Magnet without dipole coils. *Yellow:* Effect of dipole coils. Labels written in boldface refer to axes that influence the gyrotron interaction directly.

The first question will be answered in Section 3.3.2, the second question in the subsequent Sections. As already mentioned above, only shifts need to be considered concerning the insert misalignment. This allows one to keep considerations restricted to the two dimensions of the cross-section of the cavity, an approach that is found widely in the literature, see e.g. [DP95, DN04, ZM11].

In addition to the static misalignment the coaxial insert vibrates due to the internal cooling. The amplitude of this dynamic misalignment can be estimated as 50 µm (extrapolation from 30 µm in [PBD+99]) and could be taken into account in interaction simulations as time-dependent static misalignment. As a first estimation and until such interaction code extensions exist, only static misalignment is considered in this work.

3.3.2 Phenomenology

Via conformal mapping the electro- or magnetostatic fields in a cavity with arbitrarily misaligned insert, e.g. the voltage between differently charged wall and insert, can be described analytically in a bipolar coordinate system. Unfortunately, the Helmholtz differential equation in bipolar coordinates is not separable in those coordinates; thus, an analytic description of the transverse field of a TE mode (or of any other nontrivial electromagnetic field) is not possible with two independent indices m and p. This corresponds to the observation that modes in extremely misaligned cavities do usually not form neat mode patterns[1], especially not patterns that resemble aligned-cavity modes. "Extremely misaligned" with respect to a given mode refers in this work to an insert that is misaligned strongly enough that it overlaps with the onset of the first maximum of that mode. See Figs. 3.8 and 3.9 for illustration.

For small enough misalignment, mode patterns converge to aligned-cavity modes and, at least for large enough caustic radii, to hollow-cavity modes with their well-defined, distinct azimuthal and radial indices. Fortunately it turns out that modes even in cavities with practically inacceptable large insert misalignment still resemble hollow-cavity modes. Figure 3.10 shows that for 1 mm misalignment, mode $\mathrm{TE}_{49,29}\{100\}$ basically retains its shape. The first, inner mode maximum is pushed away by the insert misalignment by less than 50 µm, which is smaller than the cyclotron radius of the electrons (80 µm) and around 1/6 of the theoretical beam width (286 µm). This can be explained if the mode pattern is understood as Gaussian microwave rays being reflected at the cavity walls with spread angle $\psi = 2\ \arccos(m/\chi_{mp})$: while the mode pattern is determined by the shape of the wall, the insert plays a role only if it is in the direct path of the rays (or, due to the relatively long wavelength, next to those paths). By contrast, a smooth insert pulls the first maximum slightly towards itself, opposite to

[1] as opposed to e.g. modes in rectangular, elliptical, or confocal-parabolic cavities

the misalignment. With the considerations given in Section 3.3.4 this is disadvantageous concerning electron energy spread.

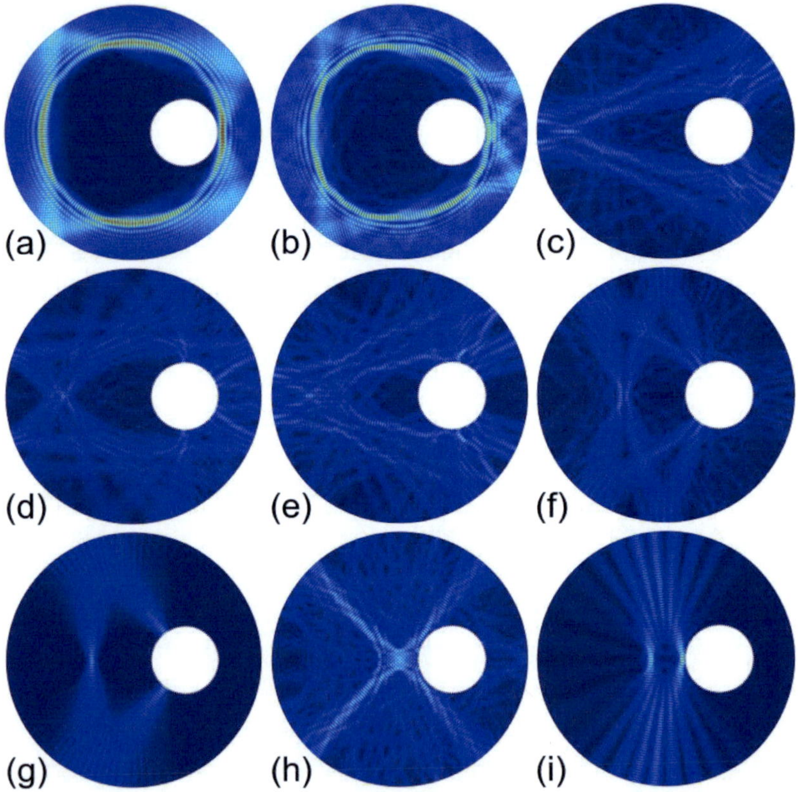

Figure 3.8: Some examples for modes in the design cavity around 238 GHz where the insert is misaligned by the extreme value of 13 mm. Modes are sorted from (a) to (i) according to where the maximum electric field touches the insert, which might correspond to the relative caustic radius of equivalent modes in aligned (or hollow) cavities. Neatly structured mode patterns of otherwise heavily disturbed modes, such as (g) and (i), seem to result from a beneficial relation between cavity geometry and wavelength. The misalignment has been chosen such that Graf's theorem is not applicable; however, modes with caustic radii larger than mode (a) resemble regular hollow-cavity modes, indicating that the non-applicability of Graf's theorem is not a sufficient condition for irregular mode shapes.

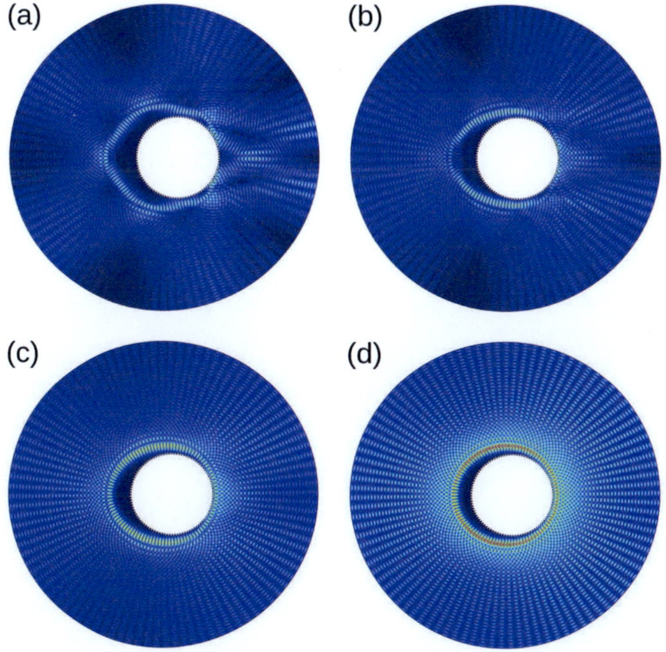

Figure 3.9: Mode $TE_{49,29}\{100\}$ in a coaxial cavity with misalignment of (a) 3.0 mm (239.130 GHz), (b) 2.5 mm (238.390 GHz), (c) 2.0 mm (237.850 GHz), and (d) 1.5 mm (237.540 GHz). With decreasing misalignment the mode pattern becomes more regular.

In analogy to the considerations in Section 3.2 one could describe the misaligned mode pattern in terms of Fourier components of the complex electric field along the nominal electron beam radius (Eq. 3.2). However, this approach has been used there only because the SIM was not applicable. In the following, an exact description of TE modes in misaligned cavities within the assumptions of SIM will be given, rendering the Fourier approach obsolete.

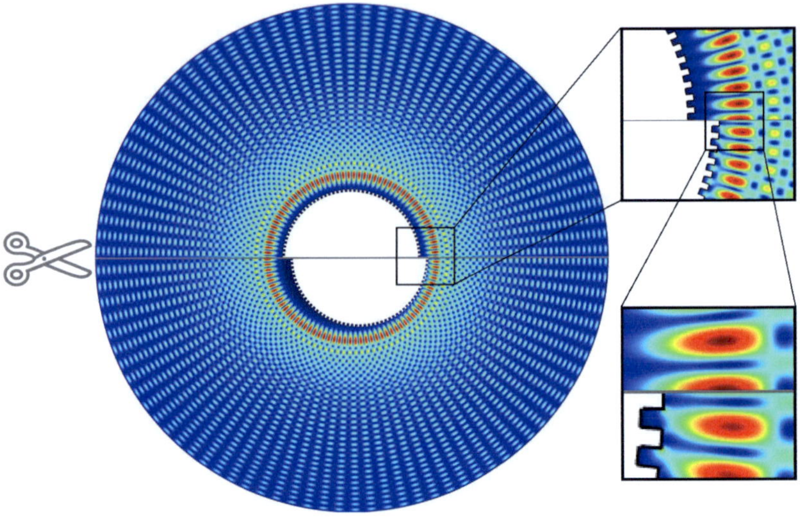

Figure 3.10: Graphical comparison of mode $\mathrm{TE}_{49,29}\{100\}$ in an aligned cavity (upper half) and in a cavity with insert misalignment of 1 mm to the right (lower half). Due to the calculation procedure both modes are slightly rotated towards each other. The boxes on the right-hand side demonstrate how tiny the effect of insert misalignment experienced by the mode maximum is.

3.3.3 Theory of Misaligned Modes

In this Section the theory of misaligned modes in cavities with corrugated insert, assuming validity of SIM, is presented. The derivations here are a logical continuation of the considerations in Annex B and follow the same formalism.

Calculation of Eigenvalue and Field Amplitudes

In order to derive the basic equations for the misaligned insert, one can follow Zhang and Thumm [ZT00]. The general expression for the B_z component expressed in the coordinate system of the inner rod $(r_{\mathrm{in}}, \varphi_{\mathrm{in}})$ using Bessel (J_m) and Neumann (N_m) functions with constant coefficients $\hat{B}_{J,m}$, $\hat{B}_{N,m}$ is

$$\hat{B}_z(r_{\text{in}}, \varphi_{\text{in}}) = \sum_m \left(\hat{B}_{J,m} J_m(k_\perp r_{\text{in}}) + \hat{B}_{N,m} N_m(k_\perp r_{\text{in}})\right) e^{-\mathrm{i}m\varphi_{\text{in}}}$$

Using Graf's addition formula for misalignment between the axes of inner rod and outer wall D_{IO}, one can express the same field in the coordinates of the outer wall (r, φ):

$$\hat{B}_z(r, \varphi) = \sum_m \sum_{q=-\infty}^{\infty} \left(\hat{B}_{J,m} J_q(k_\perp r) + \hat{B}_{N,m} N_q(k_\perp r)\right) J_{q-m}(k_\perp D_{\text{IO}}) e^{-\mathrm{i}q\varphi} \tag{3.3}$$

Graf's theorem is valid for Neumann functions only if $D_{\text{IO}} < r$ because the pole of the Neumann functions at $r_{\text{in}} = 0$ cannot be approximated by functions which are smooth at this point. This inequality can only be violated if the misalignment is at least half the insert radius, $2D_{\text{IO}} \geq R_{\text{I}}$, which is not considered for physical reasons anyways. The impedance boundary condition at the cavity wall in the aligned geometry as expressed in Eq. B.11 leads to

$$Z_{\text{O}} = \frac{\mathrm{i}\mu_0\omega}{k_\perp} \cdot \frac{\sum_m \sum_{q=-\infty}^{\infty} \left(\hat{B}_{J,m} J'_q(k_\perp R_{\text{O}}) + \hat{B}_{N,m} N'_q(k_\perp R_{\text{O}})\right) J_{q-m}(k_\perp D_{\text{IO}}) e^{-\mathrm{i}q\varphi}}{\sum_m \sum_{r=-\infty}^{\infty} \left(\hat{B}_{J,m} J_r(k_\perp R_{\text{O}}) + \hat{B}_{N,m} N_r(k_\perp R_{\text{O}})\right) J_{r-m}(k_\perp D_{\text{IO}}) e^{-\mathrm{i}r\varphi}} \tag{3.4}$$

for the misaligned insert. The following considerations will be restricted to a perfectly conducting outer wall, with impedance $Z_{\text{O}} = 0$.

Since the Eq. 3.4 has to be true for all φ, one obtains:

$$\sum_m \left(\hat{B}_{J,m} J'_q(k_\perp R_\mathrm{O}) + \hat{B}_{N,m} N'_q(k_\perp R_\mathrm{O})\right) J_{q-m}(k_\perp D_\mathrm{IO}) = 0 \quad \forall q \in Z \tag{3.5}$$

which has to be solved for unknown field amplitudes $\hat{B}_m$ (that replace $\hat{B}_{J,m}$ and $\hat{B}_{N,m}$, see Eq. B.13) and for eigenvalue $\chi_\mathrm{O} \equiv k_\perp R_\mathrm{O}$. With the definition of the characteristic matrix

$$K_{qm}(\chi) := J_{q-m}\left(\chi_\mathrm{O}\frac{D_\mathrm{IO}}{R_\mathrm{O}}\right) \cdot \left(J_q(\chi) \quad , \quad -N_q(\chi)\right) \cdot \\ \cdot \begin{pmatrix} N'_m\left(\frac{\chi_\mathrm{O}}{C_R}\right) & N_m\left(\frac{\chi_\mathrm{O}}{C_R}\right) \\ J'_m\left(\frac{\chi_\mathrm{O}}{C_R}\right) & J_m\left(\frac{\chi_\mathrm{O}}{C_R}\right) \end{pmatrix} \cdot \begin{pmatrix} \frac{2\pi}{C_R}\cos\left(\frac{\chi_\mathrm{O} D_{r,\mathrm{I}}}{R_\mathrm{O}}\right) \\ \frac{M_\mathrm{I} D_{\varphi,\mathrm{I}}}{R_\mathrm{O}}\sin\left(\frac{\chi_\mathrm{O} D_{r,\mathrm{I}}}{R_\mathrm{O}}\right) \end{pmatrix} \tag{3.6}$$

and of its derivative analogous to Eq. B.21,

$$[\partial_\chi K_{qm}(\chi)]_{\chi_\mathrm{O}} := J_{q-m}\left(\chi_\mathrm{O}\frac{D_\mathrm{IO}}{R_\mathrm{O}}\right) \cdot \left(J'_q(\chi_\mathrm{O}) \quad , \quad -N'_q(\chi_\mathrm{O})\right) \cdot \\ \cdot \begin{pmatrix} N'_m\left(\frac{\chi_\mathrm{O}}{C_R}\right) & N_m\left(\frac{\chi_\mathrm{O}}{C_R}\right) \\ J'_m\left(\frac{\chi_\mathrm{O}}{C_R}\right) & J_m\left(\frac{\chi_\mathrm{O}}{C_R}\right) \end{pmatrix} \cdot \begin{pmatrix} \frac{2\pi}{C_R}\cos\left(\frac{\chi_\mathrm{O} D_{r,\mathrm{I}}}{R_\mathrm{O}}\right) \\ \frac{M_\mathrm{I} D_{\varphi,\mathrm{I}}}{R_\mathrm{O}}\sin\left(\frac{\chi_\mathrm{O} D_{r,\mathrm{I}}}{R_\mathrm{O}}\right) \end{pmatrix} \tag{3.7}$$

one can write Eq. 3.5 as a vector equation:

$$\underline{K}' \cdot \vec{\hat{B}}_\mathrm{in} = \vec{0} \tag{3.8}$$

Here, $\vec{\hat{B}}_\mathrm{in}$ represents the field amplitudes with respect to the inner-rod coordinate system, while $\underline{K}$ relates these amplitudes to the electromagnetic field at the cavity wall, specifically as $\underline{K}'$ to the boundary condition.

For nontrivial solutions of Eq. 3.8, i.e. where $\hat{\vec{B}}_{\text{in}} \neq \vec{0}$, $\det\left(\underline{K}'\right)$ has to vanish and $\hat{\vec{B}}_{\text{in}}$ is the kernel of $\underline{K}'$. For almost all (in a mathematical sense) radii ratios C_R, and therefore in numerical investigations, all roots of $\det\left(\underline{K}'(\chi)\right)$ are single roots, thus the kernel is one-dimensional, i.e. the solution is unique up to a (real-valued) constant: the energy content of the mode. If the misalignment vanishes, $D_{\text{IO}} = 0$, $\underline{K}'$ is a diagonal matrix, the determinant of which is just the product of its diagonal elements, $\det\left(\underline{K}'(\chi)\right) \equiv \prod_m K'_{mm}(\chi)$. This product is obviously zero if (at least) one of the K'_{mm} is zero, and hence it can be reduced to the well-known Eq. B.21 for all azimuthal indices m individually.

For a given maximum number of azimuthal indices m_{max} the number of roots of $\det\left(\underline{K}'(\chi)\right)$ below a given χ_{max} is roughly independent of the misalignment, which confirms the intuitive understanding that possible eigenmodes neither appear nor vanish just by moving the insert. But this also means that by counting all roots corresponding to any given m up to χ_{max}, all those misaligned-cavity eigenmodes representing "different radial indices" are included, even though m and p might completely lose the intuitive meaning they had in hollow cavities.

Azimuthal Components

The individual field components at the outer wall (expressed in the outer wall system) $\hat{B}_{q,\mathrm{O}}$ can be obtained via

$$\begin{aligned} \hat{\vec{B}}_{\mathrm{out}}(R_{\mathrm{O}}) &= \underline{K}(\chi_{\mathrm{O}}) \cdot \hat{\vec{B}}_{\mathrm{in}} \\ \iff \quad \hat{B}_q(R_{\mathrm{O}}) &= \sum_m K_{qm}(\chi_{\mathrm{O}})\hat{B}_m \quad \forall q \in Z \end{aligned} \tag{3.9}$$

Similarly, for a beam concentric to and expressed in coordinates of the outer wall, with $\chi_{\mathrm{b}} = \chi_{\mathrm{O}}/R_{\mathrm{O}} \cdot r_{\mathrm{b}}$:

$$\begin{aligned} \hat{\vec{B}}_{\mathrm{out}}(r_{\mathrm{b}}) &= \underline{K}(\chi_{\mathrm{b}}) \cdot \hat{\vec{B}}_{\mathrm{in}} \\ \iff \quad \hat{B}_q(r_{\mathrm{b}}) &= \sum_m K_{qm}(\chi_{\mathrm{b}})\hat{B}_m \quad \forall q \in Z \end{aligned} \tag{3.10}$$

If one attempts at describing the *outer* wall boundary condition in terms of the *inner* wall coordinate system, one has to identify the indices q and r in the expression for Z_{I} analogous to Eq. 3.4, which seems less straightforward. For an impedance corrugation, the respective matrix shall not vanish, but return the boundary fields for all azimuthal components. Thus, to obtain an equation of the form Eq. 3.8 that can be solved, one has to substract the impedance boundary condition from individual matrix terms. The result is simply the transposed matrix (multiplied by the impedance), $K_{mq}^{\prime(\mathrm{out}\to\mathrm{in})} = K_{qm}^{\prime(\mathrm{in}\to\mathrm{out})}$, which is comprehensible since for each root χ_{O} both matrices have to vanish simultaneously, and a transposition is the only nonarbitrary matrix transformation that ensures this property. $\hat{\vec{B}}^{\mathrm{T}}_{\mathrm{out}}$ is then the cokernel of $\underline{K}'$.

Of course, one has to express the field components at the coaxial insert in its own coordinate system, which, with field amplitudes analogous to Eq. 3.8, yields

$$\begin{aligned} \hat{\vec{B}}_{\text{in}}(R_{\text{I}}) &= \underline{K}^{\text{T}}(\chi_{\text{I}}) \cdot \hat{\vec{B}}_{\text{out}} \\ \iff \quad \hat{B}_m(R_{\text{I}}) &= \sum_q K_{mq}(\chi_{\text{I}})\hat{B}_q \quad \forall m \in Z \end{aligned} \tag{3.11}$$

Here, $\chi_{\text{I}} = \chi_{\text{O}}/C_R$. It turns out that, in numerical calculations with appropriately high eigenvalue precision, the field amplitudes are independent of the chosen coordinate system, $\vec{B}_{\text{out}} = \vec{B}_{\text{in}} =: \vec{B}$, as it should be. The azimuthal components $\hat{\vec{B}}_{\text{in}}(R_{\text{I}})$ and $\hat{\vec{B}}_{\text{out}}(R_{\text{O}})$, on the other hand, do depend on the choice of the coordinate system. Of course, it makes only sense to express the components at any radius in the proper coordinate system, i.e. $\hat{\vec{B}}_{\text{in}}(R_{\text{O}})$ is not a reasonable expression.

Mind that with the presented method, the eigenvalue χ_{O} (equivalently, the cutoff wavenumber $k_\perp$) for each solution is the same for all field components with indices q (or m). This means that an eigenmode in the misaligned cavity is *not* represented as a sum over aligned-cavity modes $\sum_{mp} a_{mp}\text{TE}_{mp}$ (which would of course have individual cutoff frequencies), but over mathematically constructed fields with various azimuthal symmetries. Those fields do not necessarily "fit" into the waveguide, i.e. they do not individually fulfill the boundary conditions, and all oscillate at the same frequency. Therefore, this kind of decomposition is not directly comparable to the decomposition of an electromagnetic field into (hollow- or coaxial-cavity) eigenmodes, as it would for example be relevant for mode conversion calculations at the end of the insert.

To emphasize this, one could for example consider a rectangular waveguide instead of the misaligned coaxial cavity. In analogy to the above situation, a chosen rectangular-waveguide mode $\text{TE}^{\square}_{MN}$ at cutoff shall – for whatever reason – be expressed as a linear combination of circular-waveguide modes $\text{TE}^{\bigcirc}_{qp}$. As this is a purely mathematical procedure, the underlying physics must remain unchanged. If the circular-waveguide modes had various cutoff frequencies, some would be below cutoff and some above, and consequently there would be dispersion along the waveguide.

This would be in contradiction to the presupposition for mode $\mathrm{TE}^{\square}_{MN}$, so the $\mathrm{TE}^{\bigcirc}_{qp}$ modes must all have the same frequency. In order to fulfill the rectangular boundary condition, the relative strengths of the $\mathrm{TE}^{\bigcirc}_{qp}$ modes must be chosen properly, and of course, none of them fulfills this rectangular boundary condition individually. It is also apparent that the radial indices p of the obtained hollow-cavity modes $\mathrm{TE}^{\bigcirc}_{qp}$ are irrelevant since the modes are only fixed by their cutoff frequency and by their relative strength, and therefore all modes $\mathrm{TE}^{\bigcirc}_{qp}$ with the same azimuthal index q are equivalent. Due to their identical cutoff frequency but different (and, because of the irrelevant p, ambiguously defined) eigenvalues, the $\mathrm{TE}^{\bigcirc}_{qp}$ do not even coexist in the same circular waveguide. To emphasize the distinction between those $\mathrm{TE}^{\bigcirc}_{qp}$ and what is usually understood as a set of "circular-waveguide modes", the $\mathrm{TE}^{\bigcirc}_{qp}$ are termed "azimuthal components" in this work.

In the above example of a rectangular waveguide it had been assumed, but not proven that an expansion of mode $\mathrm{TE}^{\square}_{MN}$ in azimuthal components is unique and converges to the given mode for a large number of components. This is different from the misaligned-cavity geometry, where the requirement for Graf's theorem, $2D_{\mathrm{IO}} \geq R_{\mathrm{I}}$, ensures these assumptions.

Normalization Constant

The normalization constant $U_{\perp}^2$ of a mode in the misaligned cavity can be calculated following Eq. B.27, where the aligned-cavity mode electric field $\vec{E}$ has to be replaced by the fields over all azimuthal components $\sum_q \vec{E}_q$. Due to the mutual orthogonality of the modes, the absolute square of the sum over the components is equivalent to a summation over the absolute squares of the individual components, which can be integrated analytically considering Eq. B.29. As already discussed, the two boundaries have to be expressed in their respective coordinate system, which results in

$$
\begin{aligned}
U_\perp^2 =& \frac{\pi c^2}{k_\perp^2} \sum_q \hat{B}_{z,q}^2(R_\mathrm{O}) \cdot (\chi_\mathrm{O}^2 - q^2) - \\
&- \frac{\pi c^2}{k_\perp^2} \sum_m \hat{B}_{z,m}^2(R_\mathrm{I}) \cdot (\chi_\mathrm{I}^2 - m^2 + \varpi)
\end{aligned}
\tag{3.12}
$$

with the field contribution within the corrugations ϖ as defined in Eq. B.19. All summands are clearly always nonnegative, as it should be for energy states. For simplification, the two sums can formally be merged, yielding

$$
\begin{aligned}
U_\perp^2 = \frac{\pi c^2}{k_\perp^2} \sum_q \Big(& \hat{B}_{z,q}^2(R_\mathrm{O}) \cdot (\chi_\mathrm{O}^2 - q^2) - \\
& - \hat{B}_{z,q}^2(R_\mathrm{I}) \cdot (\chi_\mathrm{I}^2 - q^2 + \varpi) \Big)
\end{aligned}
\tag{3.13}
$$

but here, since none of those individual summands has a physical meaning for itself, some of them can be negative, especially for large misalignment, for large insert radius, or if $k_\perp D_{r,\mathrm{I}} \to \pi/2$.

Azimuthal Index of Misaligned Modes

As already discussed in Section 3.3.2, the eigenmodes (with defined eigenvalue χ_O) of a tapered cavity with corrugated misaligned and perhaps tilted insert do not necessarily correspond to hollow-cavity modes TE_{mpl}. Especially in non-rotationally symmetric systems, one should not expect that it is generally possible to uniquely assign "axial", "azimuthal", and "radial" indices. But at least under Vlasov approximation one can by definition consider TE modes that have a well-defined axial index l and for each transversal slice at coordinate z a "transversal index" w, where $w(z) \simeq (m,p)$ in axisymmetric cavities. The latter equivalence implies a possible z-dependence of the traditional indices: $m = m(z)$, $p = p(z)$. Mind that this does not mean mode conversion, it just indicates that the same eigenmode resembles a hollow-cavity mode $\mathrm{TE}_{m_1 p_1}$ at one axial coordinate z_1 and a different hollow-cavity mode $\mathrm{TE}_{m_2 p_2}$ at another coordinate z_2.

Despite this, it is convenient to retain the hollow-cavity mode designation "TE_{mp}" as far as possible. This should at least cover geometries with small enough misalignment where one can obtain both indices by counting azimuthal and radial maxima, such as in Fig. 3.9c-d or in the lower half of Fig. 3.10, at each axial position z. (Since mode profile transitions on transverse planes are smooth, this implies that m and p remain the same along the whole cavity.) Here, hollow-cavity mode component TE_{mp} within mode TE_w is large enough such that its profile dominates over all other profiles $\mathrm{TE}_{m'p'}$. Physically speaking, component TE_{mp} has the highest energy content of all components (Eq. 3.12):

$$m_{\mathrm{dom}} := \{q \mid \hat{B}^2_{z,q}(R_{\mathrm{O}}) \cdot (\chi^2_{\mathrm{O}} - q^2) \stackrel{!}{=} \max\} \tag{3.14}$$

The corresponding radial index p_{dom} is then the p-th root where $m_{\mathrm{dom}} = m$, counting from $\chi = 0$ [FAG+16b].

Figure 3.11 shows the found roots within a certain χ-C_R region, using a matrix with $m, q = 0 \ldots 65$, for two alignments. The radii ratios are sampled according to the axial coordinate, and the dominant azimuthal component is color-coded. One observes that roots with the same dominant azimuthal index m_{dom} lie on the typical χ-C_R curves, indicating that the described assignment of azimuthal indices is consistent between different C_R. Curves with lower m_{dom} are steeper than modes with high m_{dom} and are more affected by the misalignment, which confirms that modes with smaller caustic radius are interact stronger with the insert than modes with larger caustic. Figure 3.12 shows the eigenvalue curve of modes $\mathrm{TE}_{49,29}$ and $\mathrm{TE}_{45,30}$ for several misalignments.

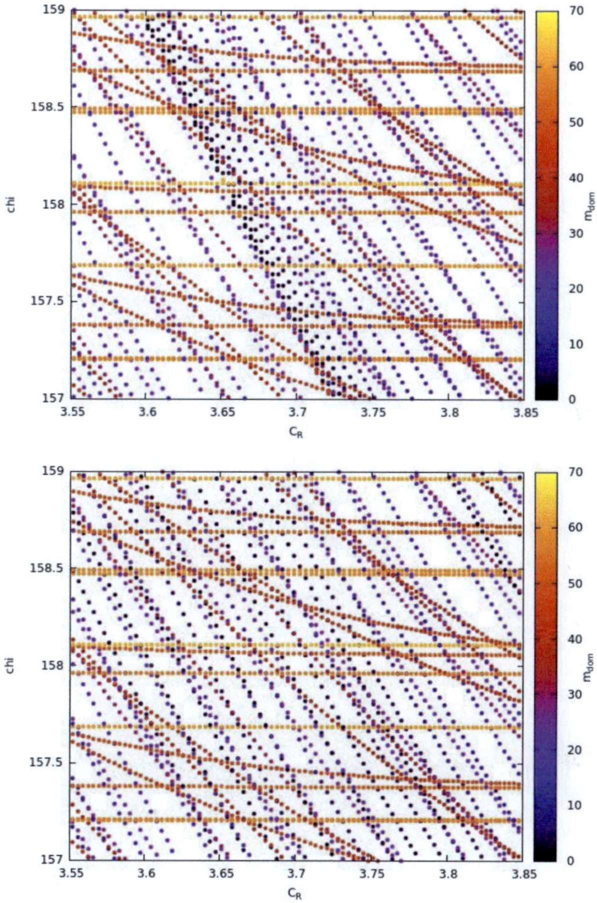

Figure 3.11: Roots of a characteristic function for $D_{\mathrm{IO}} = 0$ µm (upper) and $D_{\mathrm{IO}} = 200$ µm (lower) within a chosen χ-C_R region. The dominant azimuthal index m_{dom} is indicated for each root.

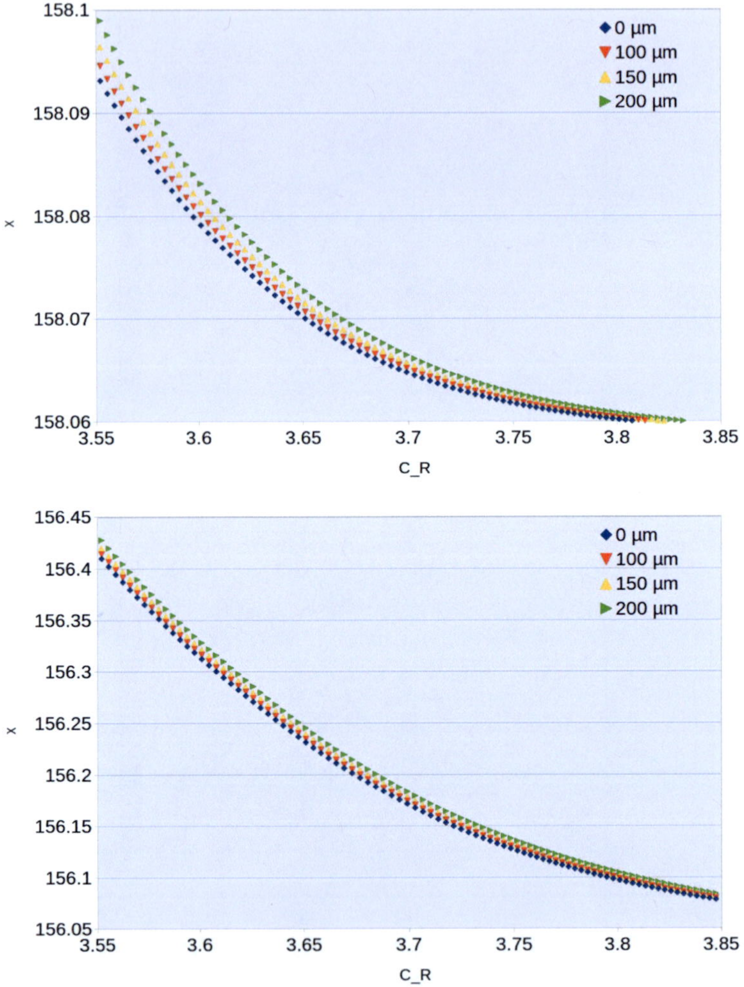

Figure 3.12: Eigenvalue curves of modes $\mathrm{TE}_{49,29}$ (upper) and $\mathrm{TE}_{45,30}$ (lower) for misalignments 0 µm, 100 µm, 150 µm, and 200 µm. Increasing misalignment as well as increasing the coaxial insert radius increases the main mode eigenvalue. Over the same C_R range, the eigenvalue of mode $\mathrm{TE}_{45,30}$ changes almost ten times more. Hollow-cavity eigenvalues are 158.058 and 156.040, respectively.

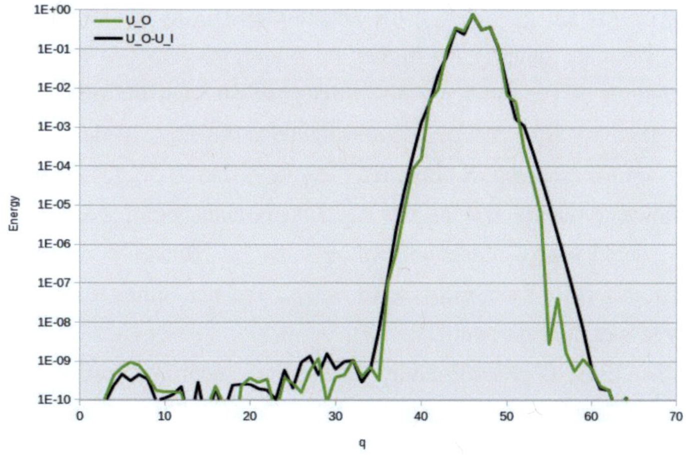

Figure 3.13: Energy content of azimuthal components q for mode $\mathrm{TE}_{46,27}$. See text for discussion.

Alternatively to Eq. 3.14 one could ask for the dominant aligned-insert component, which would additionally include the negative "I" term from Eq. 3.13. However, as already pointed out, the resulting term has no direct physical meaning and might even be less than zero. For relevant scenarios, the coaxial-cavity modes are very similar to their hollow-cavity counterparts anyways; thus, the additional term is much smaller than the term used in Eq. 3.14.

In Figure 3.13 the normalized energy spectrum of mode $\mathrm{TE}_{46,27}$ at a misalignment of 350 µm is shown: the green curve represents the energy components with respect to hollow-cavity modes, while the black curve represents the components with respect to aligned-insert modes. Both curves largely agree, especially for the largest components. One can see that the next-largest components are $q = m_{\mathrm{dom}} \pm 2$, which appears to be typical for misalignment larger than 300 µm also for the ten most relevant modes, while smaller misalignment leads to $q = m_{\mathrm{dom}} \pm 1$ as next dominant components.

As long as $m/\chi_\mathrm{O} \lesssim R_\mathrm{I}/R_\mathrm{O}$, the amplitudes $\hat{B}_m$ are all of the same order, but decrease quickly for higher m, which corresponds to the relative field strength of the mode at the insert position. On the other hand, the matrix elements K'_{qm} of Eq. 3.7, and likewise the K_{qm}, generally increase quickly with increasing m above $R_\mathrm{I}/R_\mathrm{O} \cdot \chi_\mathrm{O}$ due to $N'_m(\chi_\mathrm{O}/C_R)$. Thus, from the components $\hat{B}_q(R_\mathrm{O})$ in Eq. 3.9 one can obtain a dominant term $\hat{B}_{m_{\mathrm{dom},B}}(R_\mathrm{O})$ with a reasonable value $m_{\mathrm{dom},B}$. However, one cannot always directly infer from $m_{\mathrm{dom},B}$ to m_dom, and the values for $m_{\mathrm{dom},B}$ are slightly less consistent over C_R.

For the $\mathrm{TE}_{49,29}$ coaxial-cavity midsection described in this work, assignment of the dominant azimuthal component as azimuthal index usually works for misalignment up to 1000 µm. However, as soon as up- and downtaper are included, the dominant azimuthal index varies along the cavity axis for misalignment larger than around 400 µm. Usually, one of the strong components visible in Fig. 3.13 takes over, leading to azimuthal index jumps of $\Delta m_\mathrm{dom} = \pm 2$. This does not necessarily mean that gyrotron operation would become unstable at this point, but it indicates that interaction simulations based on modes with defined azimuthal and radial indices cannot be trusted.

Fraction of Unwanted Mode Content

Even under the assumption that only the desired operating mode is excited in the (misaligned coaxial) cavity and that it is guided through the nonlinear uptaper into the launcher without any mode conversion, it will transform into a hollow-cavity mode $\mathrm{TE}_{m_\mathrm{dom},p_\mathrm{dom}}$ as soon as the coaxial insert ends (see Fig. 1.11). Therefore, the expected stray radiation level is at least the unwanted amount of energy that is not contained in the dominant hollow-cavity component:

$$C_{\text{stray}} = 1 - \frac{\frac{\pi c^2}{k_\perp^2} \hat{B}^2_{z,m_{\text{dom}}}(R_{\text{O}}) \cdot \left(\chi_{\text{O}}^2 - m_{\text{dom}}^2\right)}{U_\perp^2} \tag{3.15}$$

This principle is schematically shown in Fig. 3.14.

The center of the cavity would here be an appropriate reference axial coordinate to determine C_{stray}; however, the dependence of the unwanted mode content on the axial position is insignificant. As an example, Fig. 3.15 shows the nine strongest azimuthal components of (arbitrarily chosen) mode $\text{TE}_{46,27}$ in a cavity with 300 µm misalignment, i.e. slightly smaller than in Fig. 3.13. The dependence of the components on the local cutoff eigenvalue (and thus on z) is small in absolute terms and even smaller in relative terms. In this special case, components $q = 44$, $q = 45$, $q = 47$ and $q = 48$ each have almost the same strength of around 28 % of the dominant component $m_{\text{dom}} = 46$, which thus carries less than 50 % of the mode's total power. Furthermore, one can observe that the component strength is roughly symmetric to the dominant component, i.e. $U_{m_{\text{dom}}-\Delta q} \approx U_{m_{\text{dom}}+\Delta q}$ for relevant Δq.

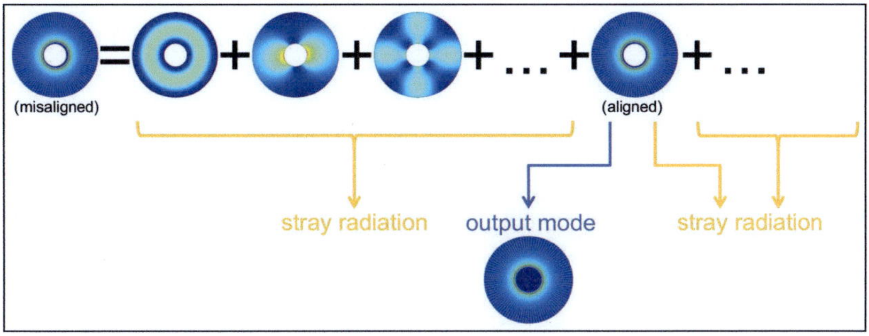

Figure 3.14: Stray radiation content of a mode in a cavity with misaligned insert. The stray radiation consists of the power carried by the non-dominant azimuthal components as well as by the part of the dominant component that differs from the corresponding hollow-cavity mode. Energy conversion from the non-dominant components into the hollow-cavity output mode does not take place because of the absence of a structure that could change the azimuthal indices.

Of course, the total stray radiation in the gyrotron is higher than C_{stray}, as it results not only from the fact that the dominant mode component has in fact a wavevector (slightly) different from that of the corresponding hollow-cavity mode, but also from spurious modes, mode conversion in uptaper and launcher, diffusive reflections from the mirrors, and reflections from the output window [RPK+10, PDD+05]. In the following it is assumed that, if a MSDC is used, the RF power lost by stray radiation as such will not decrease the gyrotron efficiency (as defined by the ratio of radiation output and electric input of the tube) below the desired requirement of e.g. 60 %. However, one would assume that the accumulated radiation in the mirror box can be reflected back into the cavity – not necessarily in opposite direction of the operating mode, but as waves with all possible k_z components –

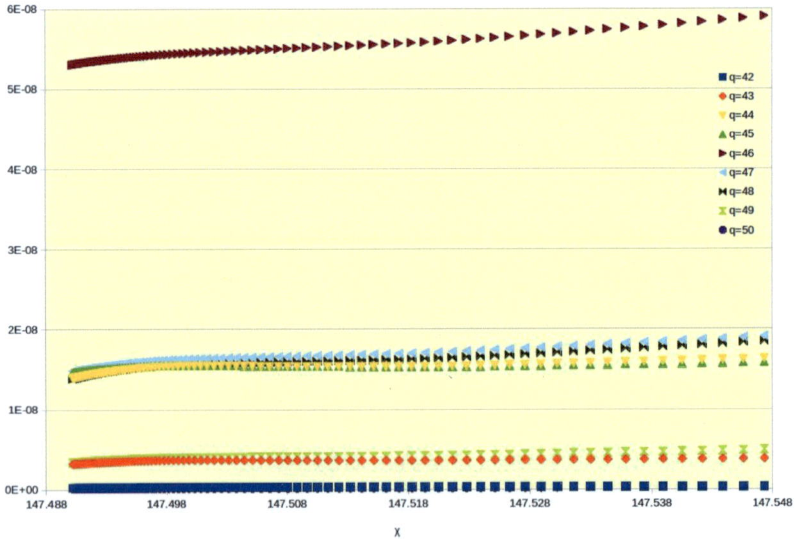

Figure 3.15: Energy content of nine azimuthal components for mode $\mathrm{TE}_{46,27}$ versus eigenvalue χ (arbitrary units). The inhomogeneous sampling results from the nonlinear dependence of χ on z. See text.

and disturb the interaction if its power relative to the generated RF power is significant. If an upper limit for internal stray radiation is defined and the basic stray radiation (with an aligned insert) is estimated, one can conclude the maximum permitted insert misalignment from Eq. 3.15.

Table 3.2 gives power estimations for a 2 MW DEMO gyrotron with a MSDC. Values have been chosen such that typical requirements on powers and efficiencies are met. In this example, electron beam generation consumes 7225 kW ($P_1 + ... + P_8$), while the auxiliary systems need 15 kW ($P_9 + P_{10} + P_{11}$), totalling 7240 kW. The beam power distributes over microwave radiation outside of the window (2050 kW, $P_3 + P_4$), residual electron energy (5000 kW, $P_1 + P_2$), and internal losses (175 kW, $P_5 + ... + P_8$). The internal losses can be apportioned into ohmic losses (65 kW, $P_7 + P_8$) and stray radiation (110 kW, $P_5 + P_6$).

From these considerations one can conclude that the stray level C_{stray} should not exceed 3.3 % in order to keep the total stray radiation in the gyrotron below 5 % of the converted electron energy. A less conservative approach would allow total stray losses of 8 % ($P_5 = 145$ kW), which results in $C_{\mathrm{stray}} \leq 6.6$ %. Even for reduced basic stray radiation $P_6 = 30$ kW, C_{stray} should not exceed 7 % to keep total stray losses below 8 %. Of course, the exact target values for stray radiation depend on the chosen power and efficiency estimations, but they do not differ considerably if other (realistic) values $P_1, ..., P_{11}$ are chosen.

Table 3.3 shows the stray levels C_{stray} in percent of the ten most relevant modes (see Table 1.1) over the cavity midsection for misalignment from 100 µm to 200 µm. In the aligned case, coaxial-cavity and hollow-cavity modes basically coincide, and the stray level is negligible. For a misalignment of 200 µm, stray radiation exceeds 7 % of the total RF energy. Stray radiation and its variation depend mostly on the mode eigenvalue for small misalignment, and increasingly on the relative caustic radius of the respective mode for larger misalignment. It increases strongly with increasing misalignment, corresponding to the exponential inner onset of the TE mode

Table 3.2: Power and efficiency estimations for a 2 MW gyrotron with MSDC.

Partial power	**Variable**	**Value (kW)**
Electron energy recovered by MSDC	P_1	4000
MSDC bombardment/heating	P_2	1000
RF outside window: Gaussian content	P_3	**2000**
RF outside window: Non-Gaussian	P_4	50
Stray radiation due to misalignment	P_5	70
Other internal stray radiation	P_6	40
Ohmic loading on the insert	P_7	5
Ohmic loading on other components	P_8	60
Operation of cooling systems	P_9	5
Operation of control systems/diagnostics	P_{10}	5
Operation of the magnet	P_{11}	5
Efficiency or share	**Formula**	**Value (%)**
Total system efficiency	$\frac{P_3}{P_2+\ldots+P_{11}}$	61.7
Gyrotron efficiency	$\frac{P_3+P_4}{P_2+\ldots+P_8}$	63.6
Interaction efficiency	$\frac{P_3+P_4+P_5}{P_1+\ldots+P_8}$	29.3
Collector efficiency	$\frac{P_1}{P_1+P_2}$	80.0
Gaussian mode content	$\frac{P_3}{P_3+P_4}$	97.6
Internal losses	$\frac{P_5+\ldots+P_8}{P_3+\ldots+P_8}$	7.9
Ohmic losses	$\frac{P_7+P_8}{P_3+\ldots+P_8}$	2.9
Stray losses (internal radiation)	$\frac{P_5+P_6}{P_3+\ldots+P_8}$	4.9
Stray level C_{stray}	$\frac{P_5}{P_3+P_4+P_5}$	3.3

fields. For any misalignment, the stray level is roughly the same for the design mode and for its main competitors; therefore, it suffices to state that a certain misalignment induces a certain amount of stray radiation. Taking the above considerations into account, insert misalignment should not exceed 150 µm for the 238 GHz 2 MW gyrotron to operate properly, and should remain below 200 µm even if comparably large stray radiation can be tolerated.

Table 3.3: Stray radiation levels C_{stray} in percent for the ten most relevant modes and for three misalignments. Minimum and maximum stray level values were evaluated under consideration of the entire cavity midsection.

Mode	100 µm	150 µm	200 µm
$\mathrm{TE}_{45,30}$	1.16 – 1.17	3.17 – 3.19	7.2 – 7.3
$\mathrm{TE}_{46,30}$	1.22 – 1.22	3.31 – 3.34	7.6 – 7.6
$\mathrm{TE}_{47,30}$	1.27 – 1.28	3.47 – 3.50	7.9 – 8.0
$\mathrm{TE}_{48,29}$	1.32 – 1.33	3.56 – 3.57	8.0 – 8.1
$\mathrm{TE}_{49,27}$	1.35 – 1.35	3.53 – 3.55	7.5 – 7.8
$\mathrm{TE}_{49,28}$	1.36 – 1.37	3.62 – 3.63	7.8 – 8.1
$\mathrm{TE}_{49,29}$	1.38 – 1.38	3.71 – 3.72	8.1 – 8.4
$\mathrm{TE}_{49,30}$	1.39 – 1.40	3.79 – 3.81	8.4 – 8.7
$\mathrm{TE}_{49,31}$	1.40 – 1.41	3.87 – 3.90	8.7 – 9.1
$\mathrm{TE}_{50,29}$	1.43 – 1.44	3.86 – 3.87	8.7 – 8.7

Ohmic Loading and Quality Factor

The total ohmic loading on a surface with radius R by a mode in a misaligned cavity is just the sum over the loadings by its individual azimuthal components:

$$w_\Omega(R) = \frac{\pi c}{2\mu_0}\frac{D_\mathrm{S}}{\lambda}\sum_m \left(\Gamma(z)^2 + \left(\frac{\Gamma'(z)}{k_\perp}\cdot\frac{m}{k_\perp R}\right)^2\right)\cdot \hat{B}^2_{\parallel,m}(R) \quad (3.16)$$

Here, D_S is the skin depth of the surface and $\Gamma(z)$ is the axial mode profile. In full analogy to the formula for the ohmic Q factor in the aligned cavity, Eq. B.32, one can express the ohmic quality factor in the misaligned geometry by sums over the azimuthal components:

$$Q_\Omega = \frac{R_\mathrm{O}}{\chi_\mathrm{O}}\cdot \\ \cdot\frac{\sum_m \left(\hat{B}^2_{z,m}(R_\mathrm{O})\left(\chi_\mathrm{O}^2 - m^2\right) - \hat{B}^2_{z,m}(R_\mathrm{I})\left(\chi_\mathrm{I}^2 - m^2 + \varpi\right)\right)}{\sum_m \left(D_\mathrm{S,O}\,\chi_\mathrm{O}\,\hat{B}^2_{z,m}(R_\mathrm{O}) + D_\mathrm{S,I}\,\chi_\mathrm{I}\,\hat{B}^2_{z,m}(R_\mathrm{I})\right)} \quad (3.17)$$

Coupling Factor

In order to determine the coupling factor[2] for misaligned fields one has to use the summation over azimuthal components (Eq. 3.9) before expanding the field in a series of cyclotron harmonics s (as in the aligned case in Eq. B.47). Clearly it is the characteristic matrix $\underline{K}$ that has to be expanded here, as it is the r-dependent quantity. Nevertheless, since $\underline{K}$ is a complicated, however linear combination of Bessel and Neumann functions, Graf's theorem is still applicable (for each m separately). Thus,

[2] also known as *coupling constant* or *coupling coefficient*

$$\hat{B}_{\parallel,q-1}(k_{qp}r)e^{\mathrm{i}(1-q)(\varphi-\varphi_{\mathrm{b}})} = \sum_m \hat{B}_m K_{q-1,m}(k_{qp}r)e^{\mathrm{i}(1-q)(\varphi-\varphi_{\mathrm{b}})} =$$
$$= \sum_m \hat{B}_m \sum_{(s-1)=-\infty}^{+\infty} K_{q-s,m}(k_{qp}r_{\mathrm{b}})J_{s-1}(k_{qp}r_{\mathrm{C}})e^{\mathrm{i}(s-1)(\pi-\varphi_{\mathrm{C}})} \tag{3.18}$$

In the expression for the coupling factor (Eq. B.50), the expressions for eigenvalue, wavevector, and frequency can be simplified to cR_{O}, which, together with $J_{s-1}(k_{qp}r_{\mathrm{C}})$, is constant for all azimuthal components of the same harmonic s. For clarity, the total field energy (Eq. 3.13) can be expressed by the sum of its components. Therefore, the coupling factor is

$$G_{qps}(r_{\mathrm{b}}) = cR_{\mathrm{O}} \cdot \frac{\hat{B}_{q-s}(r_{\mathrm{b}})}{\sqrt{\sum_m U^2_{\perp,mp}}} \cdot J_{s-1}(k_{qp}r_{\mathrm{C}}) \tag{3.19}$$

Here, Eq. 3.10 applies, using $\hat{B}_{q-s}(r_{\mathrm{b}}) = \sum_m \hat{B}_m K_{q-s,m}(k_{qp}r_{\mathrm{b}})$, as in Eq. 3.18. Clearly, this coupling factor is not well-defined because the azimuthal index q of a mode in a misaligned cavity is neither, which itself is a consequence of the underlying requirement that the coupling depends only on the radial – and not on the azimuthal – coordinate (see Eq. B.43). However, for not too large misalignment one would expect that the diagonal elements of $\underline{K}$ are substantially larger than the nondiagonal elements, thus one could assign $q \mapsto m_{\mathrm{dom}}$ to obtain a good approximate value for the coupling.

For aligned cavities, only $K_{q-s,q-s}$ and $U_{\perp,qp}$ are nonzero, and Eq. 3.19 simplifies to the well-known coupling factor, see Eq. B.50.

Numerical Implementation

This Section gives an overview of how the theory of misaligned inserts has been implemented in the code SCNCHIMP (see Annex C.2.4). Eight-bit precision is sufficient for all calculations; no numerical instabilities have been encountered. A typical full calculation for the cavity and frequency given in this work takes one day using a server with 16 Intel Xeon E5 dual-CPUs at 2.9 GHz. The most central question concerning numerical implementation of the theory presented here is how to treat the infinitely large matrix $\underline{K}'$. So far, the approach first presented in [DP95] is used exclusively in the literature, where $\underline{K}'$ extend over values $m = (m_0 - \Delta m)\ ...\ (m_0 + \Delta m)$ and $q = (m_0 - \Delta m)\ ...\ (m_0 + \Delta m)$ with sufficiently large Δm for a given set of interesting azimuthal indices $\{m_0\}$. This approach will be referred to as the "Small matrix approach" (SMA). In order to prove convergence for increasing Δm, one can consider normalized characteristic functions,

$$K'_{\text{small}}(m_0, \Delta m, \chi) := \frac{\det\left(\underline{K}'(m_0, \Delta m, \chi)\right)}{\det\left(\underline{K}'_{(m_0,m_0)}(m_0, \Delta m, \chi)\right)}$$

where $\underline{K}'_{(m_0,m_0)}$ denotes the (m_0, m_0)-minor of $\underline{K}'$.

However, during calculation of eigenvalues for m_0 in the misaligned case, one also obtains spurious roots (and, in the normalized case, poles) due to the higher-oder matrix determinant. Those modes are spurious insofar as they correspond to azimuthal indices $m \neq m_0$ which are the central indices of other Small matrices and are determined there with the desired high precision.

For large, but reasonable misalignment and for very high-order eigenvalues, the required matrix size exceeds 30 times 30 entries, which corresponds to a high density of spurious eigenvalues. For several, adjacent central azimuthal indices m_0, many of those matrices have identical entries. Therefore, combining these small matrices into one single matrix with $m = 0 \ldots (m_{\max} + \Delta m)$ and $q = 0 \ldots (m_{\max} + \Delta m)$

- reduces the total number of matrix elements to be calculated,
- enables re-using results from time-consuming Bessel and Neumann function calculations,
- gives every root a physical meaning, regardless of how precisely roots for $m \to 0$ or $m \to m_{\max} + \Delta m$ are determined, and
- reduces the number of matrix determinants to be calculated.

This approach will be termed "Large matrix approach" (LMA) and will be discussed from now on. Figure 3.16 illustrates the relation between SMA and LMA. One major disadvantage of the LMA is the fact that the matrix determinant can become very large, especially for small m. However, this can be mitigated by normalizing each matrix element to the $(m_{\max}+\Delta m+1)$-th root of the absolute value of the product of the matrix's diagonal elements before subsequent calculations. One could also consider a full Large matrix, i.e. one that includes elements outside the band $\pm\Delta m$ around the main diagonal, since the corresponding Bessel and Neumann functions have to be calculated anyways. However, the results – eigenvalue and field amplitudes – will not change because the matrix elements decrease by $J_n(k_\perp D_{\mathrm{IO}})$ with distance n from the main diagonal and are therefore typically very close to zero for large n.

For any geometry and pre-defined value $\Delta\chi$, two subsequent roots χ_{w_1}, χ_{w_2} of $\underline{K}'(\chi)$ can be closer to each other than $\Delta\chi$, and there is no method to predict the distance $\chi_{w_2} - \chi_{w_1}$. However, statistically there is a typical minimum distance between subsequent roots, which can serve as the maximum step size for a root search to find all eigenvalues with high probability. As an example, for $m_0 = 49$ and $\Delta m = 26$, $\Delta\chi \approx 5 \cdot 10^{-6}$. After one root is found, it should be resolved up to a precision of $\delta\chi \approx 10^{-9}$ to obtain correct results in subsequent calculations.

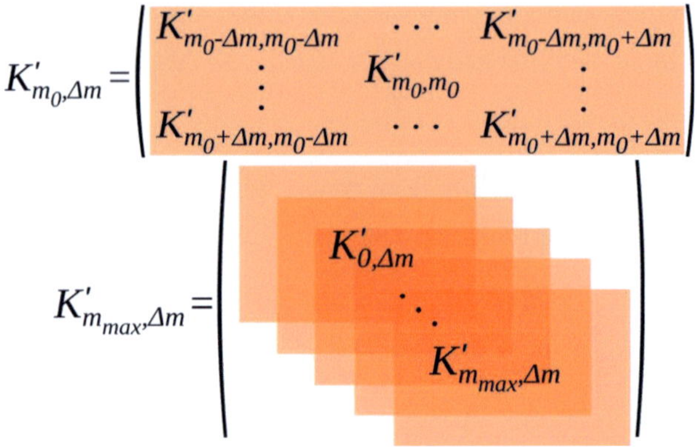

Figure 3.16: Illustration of the Small matrix approach (upper) versus the Large matrix approach (lower). Orange areas represent nonzero matrix elements. As one can see, the overlap of identical Small-matrix elements in the LMA is large. Since at least one element per row and per column in the Large matrix has to be determined anyway, an efficient algorithm allows to calculate all Large-matrix elements (including the ones outside of the Δm band) without additional need for Bessel and Neumann function calculations.

In order to analyze the bahavior of relevant modes one would usually consider a rectangle in the C_R-χ plane for each misalignment offset, where the dependence on C_R might include the effect of an insert tilt since both radii ratio and tilt-induced misalignment are functions of z. In both cases, SMA and LMA, root search can be efficiently parallelized over the axial coordinate sampling i_z.

Since one would start at a certain $\chi_{\min}$, one needs a (z-dependent) offset of radial indices. Additionally, due to the numerical procedure, it happens that roots e.g. of modes with $m \lesssim m_{\max} + \Delta m$ show in fact dominant indices $m_{\mathrm{dom}} \neq m$. In many, but not all cases, those modes have a large caustic radius, which results in eigenvalues that are independent of i_z and can therefore be filtered out automatically.

Otherwise this filtering has to be done manually, e.g. by checking whether roots occur at one i_z while no roots with the same m and with similar eigenvalues are present at i_{z-1} and i_{z+1}. However, it could be by coincidence that a proper root at, say, i_{z+1} has not been found because it is extremely close to another proper root, either corresponding to an inner mode (see Section 3.4) or to a mode with different azimuthal index m'. Then, i_{z+2} has also to be considered, where, however, the spurious mode might appear again. Such semi-regular patterns are relatively easy for a human to recognize, but very difficult to code.

Typically accepted values for the misalignment are $k_\perp D_{\mathrm{IO}} \lesssim 1.25$ (or $D_{\mathrm{IO}}/\lambda_\perp \lesssim 0.2$) [DP95], which would be 0.25 mm for a 240 GHz gyrotron. Qin *et al.* calculated a drop in efficiency from around 50 % to 45 % for a misalignment of $\lambda/8$ for both smooth [QYL$^+$15] and corrugated [QLY$^+$14] inserts, with the beam aligned to the cavity wall. For the 238 GHz gyrotron this critical limit corresponds to 160 µm, which coincides with the results from the above stray level investigations.

3.3.4 Voltage Depression

In addition to the mode deformation described above, misalignment of insert, beam, and wall versus each other also influences the voltage depression on the electron beam, which changes the energy distribution of the electrons and therefore affects gyrotron operation. For the same reasons as above, two-dimensional analyses are sufficient to obtain reliable results.

In order to calculate the depression on an arbitrarily positioned and shaped beam in hollow cavities or in coaxial cavities with arbitrary misalignment, the novel code WickedQueen has been developed by the author, as published in [FAG$^+$16a]. The used formulas are solely based on the method of image charges (mirror charges) [Smy68] in the less known case of geometries with translational symmetry. The electron beam is hereby described as a bundle of line-charges or, in the case of axisymmetric homogeneous beams, as one effective line-charge using the known analytic formulas [DK81]. For hollow cavities, one mirror charge per original charge is induced, while for coaxial cavities, each original charge induces infinitely many mirror charges. However, the series of mirror charges converges to the two geometrical poles of the cavity geometry which allows proper truncation of this series. Due to the superposition principle, voltage depression at each point within the electron beam (and, in fact, within the whole cavity) is then the resulting potential of all line-charges. In Fig. 3.17, this convergence can be seen. Figure 3.18 displays a thick misaligned beam in an aligned cavity. Also for geometries with more realistic beam thickness and misalignments, there are typically many electrons with high depression on the side away from the insert ("HD"), equally many electrons with low depression on the side near the insert ("LD") and some electrons very close to the insert with very low depression ("VLD").

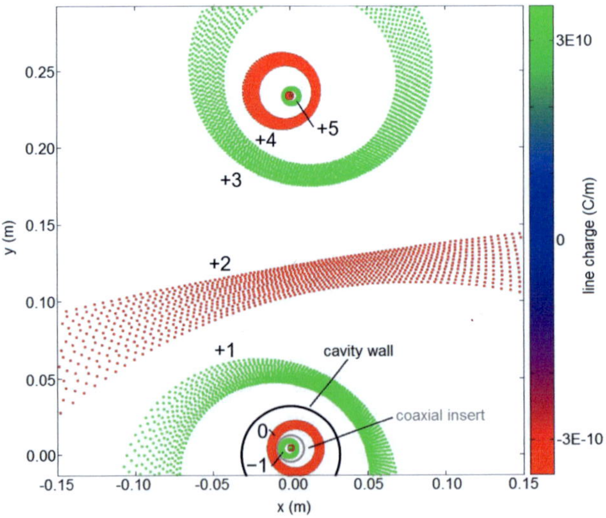

Figure 3.17: Original ("0") and mirror charges of n-th degree of a thick misaligned beam in a misaligned coaxial cavity. The signs of the charges are color-coded and alternate every iteration. Negative n represent mirror charges within the insert, while positive n represent those outside of the cavity wall.

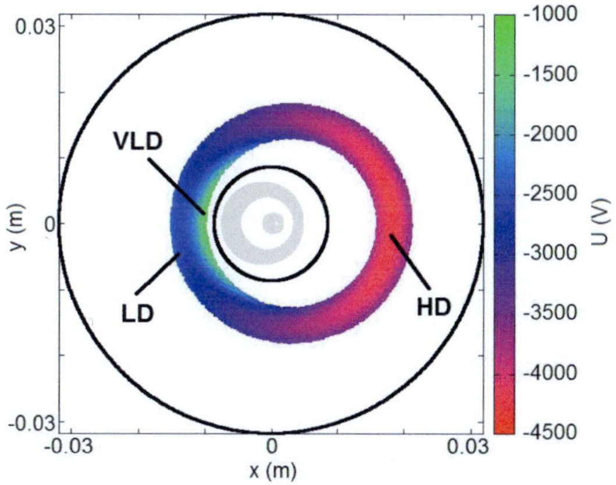

Figure 3.18: Voltage depression within a thick misaligned beam in an aligned cavity; see text.

As already mentioned, the electron beam should be placed at the mode maximum to ensure optimum interaction efficiency. Since the mode maximum does hardly move if the insert is shifted versus the cavity wall by a reasonable amount D_{IO}, one can safely assume that the electron beam has to remain in place for all D_{IO}. Figure 3.19 shows voltage depression distributions for the $\mathrm{TE}_{49,29}$ set-up and insert misalignment up to 1 mm (see also [FAG^{+}16a], fig. 5). 20011 electrons have been sorted into 50 voltage bins for each misalignment. In all cases, a VLD tail exists, covering around 200 V. For all misalignment larger than 200 µm one can distinguish the statistical HD and LD peaks of similar size. Between those peaks the voltage depression distribution is basically flat, with relative shares only slightly below 2 %, corresponding to the moderate HD and LD peaks. Solely for perfect alignment only one large statistical peak exists. As a good approximation (for larger misalignment), the width of the voltage depression distribution is basically flat and grows linearly with misalignment, in the shown case by 2 V per micrometer misalignment. A misalignment of 1 mm implies 1.2 keV peak-to-peak RMS spread, or 1.4 % of the total beam energy.

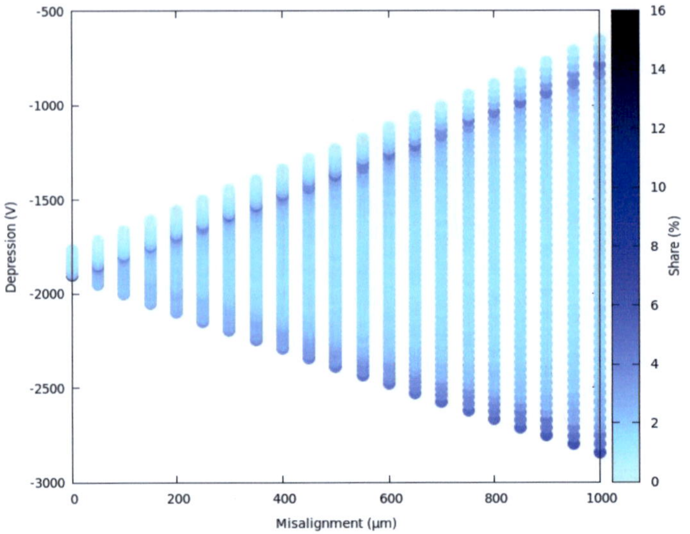

Figure 3.19: Voltage depression distributions for several misalignments.

3.4 Occurrence of Inner Modes

Here and in the following, considerations are restricted to typical coaxial high-power gyrotron geometries, where $C_R \sim 3$ and $C_\mathrm{K} \sim 1/C_R$. In Annex B.4 and the citations therein it is mentioned that the corrugations cause additional modes to form, the so-called "inner modes". It is known that those modes have a high field density at or within the corrugations and therefore impose a high ohmic loading on the insert, which strongly decreases their overall quality factor. Therefore they are not suitable as operating modes. However, these inner modes – as defined eigenmodes of the electromagnetic field in the cavity – are as valid as the normal "outer" modes and can have a comparable coupling to the electron beam. Any exclusion of them as competing modes in interaction simulations, such as EURIDICE, should therefore be justified *a posteriori*, for example by comparison of quality factors.

In the χ-C_R plane, inner modes correspond to those intervals of the eigenvalue curves where $\mathrm{d}\chi^*/\mathrm{d}C_R > 0$. In Fig. 3.20, points A to D all lie in such an interval, and one would suspect that they represent inner modes. Apart from that, the detailed nature of inner modes remained somewhat obscure.

Having COMSOL as a powerful Laplace solver at hand it was possible to track down and study the inner modes in a more realistic scenario than the SIM could provide. "Track down" because in the relevant geometry the inner modes appeared around 150 GHz to 175 GHz, which was rather close to, but not exactly at the frequency predicted by the SIM. This difference is reasonable since the inner modes originate from the corrugations and are therefore dependent on the particular groove shape, which differs slightly between SIM assumptions and the used COMSOL geometry. Figure 3.21 shows four inner modes with constant radial index $p^* = 18$ and with increasing azimuthal indices $m = 34, 37, 41, 45$, i.e. with decreasing C_{K}. The chosen modes roughly correspond to positions A to D in Fig. 3.20, respectively. One can observe the following properties:

- Inner modes are, in fact, real eigenmodes of the RF field and are not e.g. artificial results from assumptions within the Surface Impedance Model.
- The mode structure of inner modes is very similar to that of normal, outer modes, apart from the additional circular field maximum near the insert.
- The actual azimuthal field maxima of the additional ring lie in the same directions as the conventional field maxima, as opposed to the outer maxima which are shifted by half a period (see Fig. 1.6).
- Inner modes do not necessarily have their caustic radius or their conventional field maximum near the insert surface.

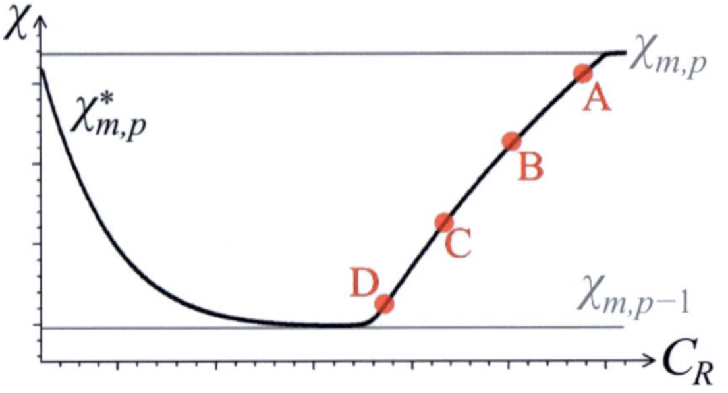

Figure 3.20: Location of four (C_R, χ) pairs A, B, C, D of a "true" inner mode (black curve). Reproduced from [Avr06], figure 2.4 (page 66), where the curve corresponds to $\mathrm{TE}_{15,7^*}$ for $D_{r,\mathrm{I}}/\lambda = 0.2$ and $C_R = 2.1 \ldots 5$.

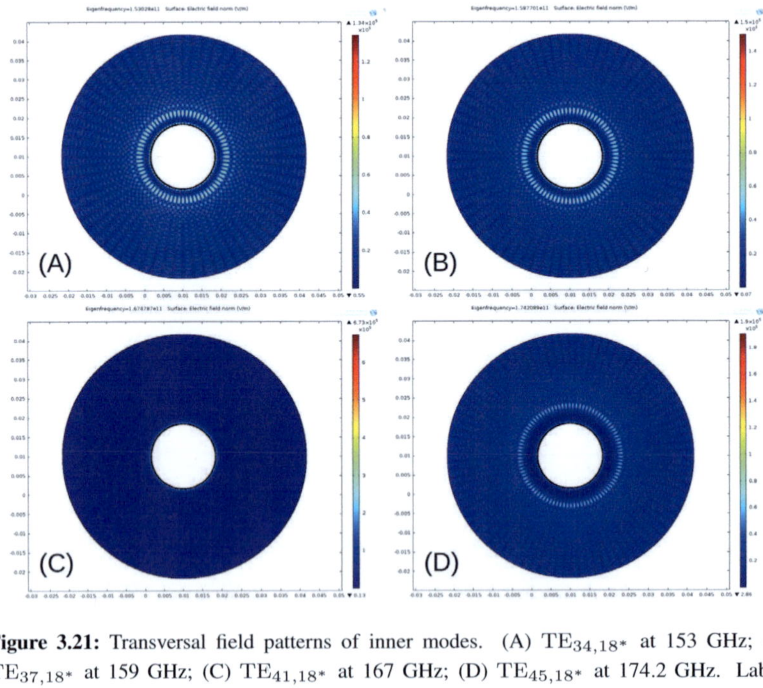

Figure 3.21: Transversal field patterns of inner modes. (A) $\mathrm{TE}_{34,18^*}$ at 153 GHz; (B) $\mathrm{TE}_{37,18^*}$ at 159 GHz; (C) $\mathrm{TE}_{41,18^*}$ at 167 GHz; (D) $\mathrm{TE}_{45,18^*}$ at 174.2 GHz. Labels A to D are roughly equivalent to position A to D in Fig. 3.20. For all four subfigures, fields are scaled to the maximum occuring electric field strength, i.e. (C) indicates that the fields inside the corrugations are very strong. See text.

- The additional field maximum is much stronger than the conventional maximum, especially for inner modes that have eigenvalues well separated from their outer-mode neighbors.

From these observations one can conclude that inner modes ν_q can indeed be interpreted as an additional set of modes which are caused by and located predominantly within the corrugations of the insert. In cases where they have eigenvalues close to an outer mode $\chi_{m,p}$, both modes couple to form a pair of hybrid modes: one with a decreased eigenvalue $\chi^*_{m,p}$, and one with an increased eigenvalue $\chi^*_{m,p+q}$.

3.5 Cavity Expansion and Deformation During Operation

Regardless of whether the cavity wall is made of pure copper, thermally hard copper alloy (e.g. Glidcop), or other highly thermal and electrical conductive materials, ohmic heating will lead to expansion of the cavity. Assuming a typical thermal expansion coefficient of copper of 17 µm/(m K), the cylindrical cavity will expand by 0.17 % (54 µm in radial direction) if heated up from 20 °C to 120 °C, and the operating frequency will be reduced by around 400 MHz. This effect is normally taken care of during the technical design phase; however, one might additionally consider the smaller cold cavity during elaborate start-up simulations.

There are no immediate reasons to expect cavity deformations during normal operation, e.g. by grossly inhomogeneous cooling or by mechanical action. However, a brief analysis of deformation effects will be made here in order to give an impression of what could be observed under such circumstances. Only first-order deformations, i.e. ellipses, are considered. As opposed to e.g. modes in misaligned cavities (see Section 3.3.2), modes in hollow cavities of elliptic cross-section can be expressed analytically via Mathieu functions [AS64], which corresponds directly to the observation that elliptic-cavity modes always have "neat" profiles, i.e. profiles that have

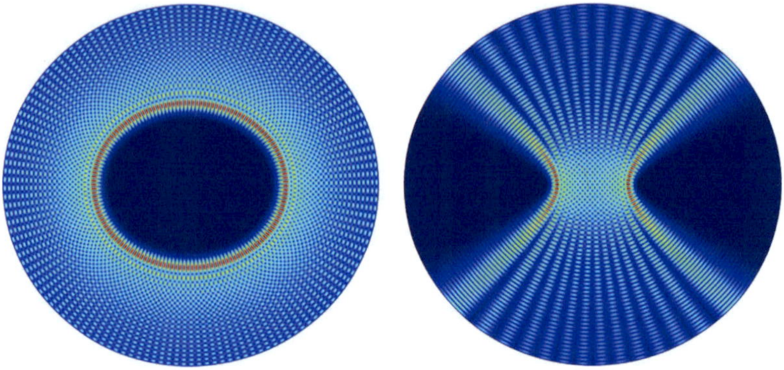

Figure 3.22: Two high-order modes in an elliptic cavity, representing the two classes of modes that were observed. See text.

periodic structures along mutually orthogonal curves (along the elliptic-hyperbolic coordinate lines of a suitable coordinate system). As shown in Fig. 3.22, there are two main classes of elliptic-cavity modes:

- Modes of the more common class look very much like asymmetric ($|m| > 0$) cylindrical-cavity modes, including clearly identifiable azimuthal and radial indices. As in the circular cavity there are always two orthogonal solutions at the same frequency. Position and elliptic shape of the first mode maximum are not significantly altered by a nearby corrugated circular insert. Their outer field maxima on the outmost ring are of equal strength, indicating that these modes would impose a rather uniform ohmic loading on the wall despite large deformation.

- Modes of the second class resemble rectangular-cavity modes which are squeezed into the elliptic boundary. They have two caustics and are of course heavily disturbed if an insert is added in the cavity center. As for axisymmetric modes ($m = 0$) in the cylindrical cavity there is only one solution per frequency, indicating a linear polarization. Due to their shape their wall loading would be inhomogeneous along the cavity circumference, with the actual peak loading position depending on the particular shape of the mode profile.

Those modes exist independently of the eccentricity of the ellipse, and therefore also in quasi-cylindrical cavities with reasonably small eccentricity. However, the second class of modes should become indistinguishable from symmetric modes. Analytic expansion of coaxial-elliptic-cavity modes into aligned-circular-cavity modes, for example in order to calculate the beam coupling, should be possible, but is of limited use.

A more quantitative analysis can be based on maximum temperature-induced deformations, as given in Table 3.4 together with the corresponding eigenfrequencies, under the assumption that such deformations are not larger than thermal expansion. With those still small deviations from the cylindrical cavity, the eigenfrequencies do not change since the cavity cross-section remains basically constant.

Figure 3.23 shows the mode patterns of scenarios No. 1 and 2 directly adjacent to each other. As one can see in the deformed case, the first (innermost) mode maximum shifts by roughly *twice* the difference in wall radius, i.e. 100 μm towards the insert in vertical direction and 100 μm away from the insert in horizontal direction. This factor of 2 depends only on the relative caustic size of the considered mode, but not on the magnitude of deformation, and decreases with increasing caustic size.

Table 3.4: Frequency of mode $\mathrm{TE}_{49,29}\{100\}$ at cutoff in cavities with various temperature-induced deformations, given as expansions/contractions in horizontal (x) and vertical (y) directions.

No.	Phenomenon/Situation	Δx (µm)	Δy (µm)	f (GHz)
1	Cold cavity	± 0	± 0	237.31
2	Deformed cold cavity (worst case)	$+54$	-54	237.31
3	Nonuniform heat-up	$+54$	± 0	237.11
4	Heated cavity	$+54$	$+54$	236.90
5	Nonuniformly heated (worst case)	$+108$	± 0	236.90

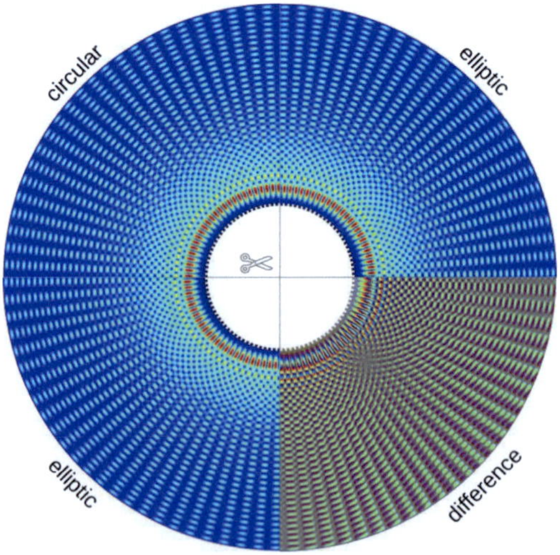

Figure 3.23: Mode $\mathrm{TE}_{49,29}$ in the perfectly circular cavity (No. 1 in Table 3.4; upper left quadrant) and in the corresponding elliptic cavity (No. 2; upper right and lower left quadrants). The lower right quadrant shows the difference in field strength between both scenarios: the greyer the area, the smaller is the difference between both fields, while saturated colors signify large differences.

4 Conclusion and Outlook

The physical design of MIG and cavity for a 2 MW 238 GHz coaxial-cavity long-pulse gyrotron has been developed. With perfectly aligned components an interaction efficiency of more than 30 % can be reached using present-day technology, considering a triode magnetron injection gun, realistic beam parameters, and a sufficient voltage margin to ensure operation stability during possible overshoots.

Operation at significantly different frequencies (multi-frequency operation), e.g. operation at 170 GHz, 204 GHz, and 238 GHz, is especially feasible for hollow-cavity gyrotrons operating at higher-order modes, provided a proper mode series selection. In coaxial-cavity gyrotrons, however, ohmic loading on the insert increases significantly with decreasing frequency, which implies either short-pulse operation only, a flexible or movable insert, or different insert dimensions for different target frequencies. None of the other investigated gyrotrons components appear to be critical in this respect.

In order to do systematic cavity and MIG optimizations in the future, it is advisable to precisely define the technical and physical boundary conditions and the design goals at an early stage and to automatize the optimization process as much as possible.

Although it seems technically possible to build a superconducting gyrotron magnet with sufficient field strength and bore-hole diameter, its development effort and manufacturing exceeds that of state-of-the-art magnets. Furthermore, the strong magnetic field and large bore-hole requires a prolonged beam tunnel and stronger magnetic fields in the collector region compared to present-day designs. Definition of a minimal set of magnet requirements instead of a detailed magnetic field profile has proven to be sufficient at an early design stage.

In order to obtain experimental and manufacturing expertise with coaxial-cavity MW-class gyrotrons at frequencies significantly above 170 GHz, it would be advisable to focus on the 200 GHz region and the highest output power that the magnet bore-hole – at best including dipole coils for beam adjustment – and and an optimized MIG design can provide.

The required number of longitudinal corrugations on the coaxial insert is relatively moderate, with approximately twice the design mode's azimuthal index being sufficient: $M_\mathrm{I} = 100$. The dimensions of the corrugations are then around 0.3 mm $\times$ 0.3 mm, which can be manufactured without difficulty. As a small, however noticeable discrepancy between SIM and FEM has been observed in the aligned geometry, the validity of SIM should be verified in misaligned geometries.

At high frequencies stray radiation increases drastically even for small misalignment of the coaxial insert, with around 150 μm misalignment corresponding to stray radiation below 4 % in the design presented in this work. In order to maximize interaction efficiency the electron beam has to remain aligned with the cavity wall, which increases the energy spread of the electrons significantly due to nonuniform voltage depression. Therefore, precise aligment of the insert in evacuated tubes is an indispensable capability.

Mathematical description and simulation of insert misalignment is possible for modes with eigenvalues up to at least 180 and for misalignment beyond the operational tolerance. With the codes WickedQueen and SCNCHIMP developed in the frame of this work, in-depth studies of misaligned cavities can be conducted, e.g. on the detailed dependence of mode eigenvalues and stray radiation on insert shift and tilt, or on mode conversion at the end of the insert. In order to run detailed time-dependent, self-consistent multimode simulations of cavities with misaligned (including vibrating) insert, the interface to interaction codes has to be completed.

Regarding the disadvantages of the coaxial insert, an in-depth comparison between multi-frequency hollow- and coaxial-cavity gyrotrons for DEMO frequencies, both theoretically and experimentally, should be carried out in the future. It might turn out that for small wavelengths the insert is not convincingly effective (or less effective than other means) for suppression of competing modes, but that its decrease of voltage depression and parasitic oscillations is indispensable. If so, a smooth insert with a radius well below the caustic radius of the operating mode with lowest frequency would be sufficient.

In order to obtain a reliable concept for a DEMO gyrotron the next critical components, inverse triode MIG and multi-stage depressed collector, need to be designed in detail. For highly efficient high-power gyrotrons emitting at small wavelengths, thin emitters that can support large current density over a long lifetime become increasingly important.

A Derivation of Equations Relevant for Gyrotrons (Classical Electrodynamics)

In Annexes A and B the gyrotron-relevant theory used in this work is derived from very first principles. This Annex A is focussed on general electrodynamic phenomena, while Annex B contains the theory of TE modes in cylindrical waveguides with aligned and corrugated coaxial insert. Main purpose of these Annexes is to provide the theoretical basis for Section 3.3.3 and for other formulas in the main part of this thesis. Most of the derivations herein have been reformulated with respect to the cited literature for the sake of intelligibility, formal exactness, and/or elegance.

As the elaboration below encompass electrodynamics from its very foundations to the description of specific phenomena it might also provide some inspiration for further theoretical investigations of gyrotron-relevant phenomena.

A.1 Physical Model Assumptions

In typical fusion gyrotrons, the electrons move in a magnetic field of the order of several teslas, at frequencies above 100 GHz. Their velocity perpendicular to the magnetic field is typically above $10\ \%$ of the vacuum speed of light c. Thus, their angular momentum is more than 10^6 times larger than their spin. Furthermore, their de-Broglie wavelength is at the nanometer scale, much smaller than the wavelength of the microwave photons, and consequently, their energy (and momentum) is by a factor 10^6 larger than the energy of the photons (which is around 1 meV). The energy of the photons, however, is considered to be much larger than the experimental upper bound for a hypothetical photon rest mass [O+14]. The number of photons in a typical high-power gyrotron cavity is at least of the order of 10^{18}. For these reasons, the gyrotron interaction may be regarded as a classical (non-quantum) interaction [Jac99]. Of course, gravitational effects can be ignored as well.

A.2 Definitions and General Relations

All coordinate systems are right-handed. The metric tensor $\eta^{\mu\nu}$ has signature $(+,-,-,-)$. Thus, raising or lowering indices leaves the sign of the scalar (temporal) component unchanged, while the three vector (spacial) components change their sign. Einstein's summation convention holds. The four-dimensional quantities are defined as follows:

- Covariant derivative (derivative in contravariant coordinates; in m^{-1}):

$$\partial_\mu = \left(\tfrac{1}{c}\partial_t \quad , \quad \nabla\right) = \eta_{\mu\nu}\begin{pmatrix}\frac{1}{c}\partial_t \\ -\nabla\end{pmatrix} = \eta_{\mu\nu}\partial^\nu$$

- D'Alembert operator (in m^{-2}): $\Box = \partial_\mu\partial^\mu = \frac{1}{c^2}{\partial_t}^2 - \Delta$

- Infinitesimal vector (differential; in m): $\mathrm{d}x^\mu = \begin{pmatrix}c\mathrm{d}t \\ \mathrm{d}\vec{x}\end{pmatrix}$

- Dimensionless momentum (4-velocity) and invariant velocity y^μ (in $\mathrm{m\,s}^{-1}$), with τ as the proper time:

$$u^\mu = \frac{\mathrm{d}x^\mu}{c\mathrm{d}\tau} = \begin{pmatrix}\gamma \\ \gamma\vec{\beta}\end{pmatrix} \qquad y^\mu = cu^\mu = \begin{pmatrix}c\gamma \\ \gamma\vec{v}\end{pmatrix}$$

- Particle current density (in $\mathrm{m}^{-2}\,\mathrm{s}^{-1}$): $\iota^\mu = ncu^\mu = \begin{pmatrix} c\varrho \\ \vec{\iota} \end{pmatrix}$

- Current density (in $\mathrm{C\,m}^{-2}\,\mathrm{s}^{-1}$) and charge density (in $\mathrm{C\,m}^{-3}$):

$$j^\mu = \mathrm{q_e} ncu^\mu = \begin{pmatrix} c\rho \\ \vec{j} \end{pmatrix} \qquad \jmath^\mu = \frac{j^\mu}{c} = \begin{pmatrix} \rho \\ \vec{j}/c \end{pmatrix}$$

  Mind that the ordinary charge density $\rho = \gamma n \mathrm{q_e}$ is not a Lorentz invariant.

- (Electron) momentum (in $\mathrm{kg\,m\,s}^{-1}$) and wavevector (in m^{-1}):

$$p^\mu = \mathrm{m_e} cu^\mu = \begin{pmatrix} W/c \\ \vec{p} \end{pmatrix} \qquad k^\mu = \frac{p^\mu}{\hbar} = \begin{pmatrix} \omega/c \\ \vec{k} \end{pmatrix}$$

- Momentum density (mass current density; in $\mathrm{kg\,m}^{-2}\,\mathrm{s}^{-1}$):

$$\pi^\mu = np^\mu = \begin{pmatrix} w/c^2 \\ \vec{\pi} \end{pmatrix}$$

- Electromagnetic potential (in $\mathrm{kg\,m\,C}^{-1}\,\mathrm{s}^{-1}$): $A^\mu = \begin{pmatrix} \Phi/c \\ \vec{A} \end{pmatrix}$

- Electromagnetic field tensor (in kg $\mathrm{C}^{-1}\,\mathrm{s}^{-1}$): $F^{\mu\nu} := \partial^\mu A^\nu - \partial^\nu A^\mu$
It follows that $F^{\mu\nu} = -F^{\nu\mu}$ and thus $F^{\mu\mu} = 0$. The definitions of the electric and magnetic fields are:

$$\vec{E} := -c\vec{e_i}F^{0i} = -c\vec{e_i}\left(\partial^0 A^i - \partial^i A^0\right) = -\partial_t \vec{A} - \nabla\Phi \tag{A.1a}$$

$$\vec{B} := \vec{e_i}\frac{\epsilon^{ijk}}{2}F_{jk} = \vec{e_i}\frac{\epsilon^{ijk}}{2}\left(\partial_j A_k - \partial_k A_j\right) \equiv \nabla \times \vec{A} \tag{A.1b}$$

Thus:

$$F^{\mu\nu} \equiv \begin{pmatrix} 0 & -E_1/c & -E_2/c & -E_3/c \\ E_1/c & 0 & -B_3 & B_2 \\ E_2/c & B_3 & 0 & -B_1 \\ E_3/c & -B_2 & B_1 & 0 \end{pmatrix} \tag{A.2}$$

Some simple known quantities can now shown to be scalars / relativistic invariants (i.e., are a product of a covariant and a contravariant vector):

- Charge conservation and continuity equation:

$$c\partial_\mu j^\mu \equiv \partial_t \rho + \nabla \cdot \vec{j} = 0 \qquad c\partial_\mu \iota^\mu \equiv \partial_t \varrho + \nabla \cdot \vec{\iota} = 0$$

- Lorenz gauge: $\partial_\mu A^\mu \equiv \frac{1}{c^2}\partial_t \Phi + \nabla \cdot \vec{A} = 0$
- Four-momentum: $p_\mu p^\mu \equiv \frac{W^2}{c^2} - \vec{p}\cdot\vec{p} = m^2 c^2$
- Relation between relativistic factor γ and speed β:

$$u_\mu u^\mu \equiv \gamma^2(1-\beta^2) = 1$$

From this, one can easily verify the time dilation $\frac{\mathrm{d}}{\mathrm{d}\tau} = \gamma\frac{\mathrm{d}}{\mathrm{d}t}$.

A.3 Maxwell's Equations and Lorentz Force

The following considerations are loosely based on [LL97], sections 23, 28 and 30. The Lagrange density of the classical electromagnetic interaction can be found to be:

$$\Lambda = -\pi_\mu y^\mu - A_\mu j^\mu - \frac{c}{4Z_0} F^{\mu\nu} F_{\mu\nu} \tag{A.3}$$

The first term is the kinetic energy of free particles, the third term describes the energy of a free electromagnetic field, and the second term describes the potential energy: the coupling between particle charges and field. All quantities are expressed as energy densities (energy per invariant volume).

The respective action is defined as the integral of the Lagrange density over the whole space and over a time interval $[t_1; t_2]$ in any inertial frame, moving at relativistic factor γ:

$$\Sigma := \int\limits_{t_1}^{t_2} \int\limits_{\mathbb{R}^3} \Lambda \mathrm{d}^3 x \mathrm{d}t = -\int\limits_{t_1}^{t_2} \int\limits_{\mathbb{R}^3} \left(\pi_\mu y^\mu + A_\mu j^\mu + \frac{c}{4Z_0} F^{\mu\nu} F_{\mu\nu} \right) \mathrm{d}^3 x \mathrm{d}t$$

The principle of stationary action now states that the action is invariant under variation of the free parameters of the system, which are the coordinates x^μ and the potentials A^μ, inside the fixed boundaries. The current density will be written as $j^\mu = n\mathrm{q_e} c u^\mu$ if necessary. The potentials themselves are explicit functions of the coordinates (the action is a functional); therefore, the chain rule applies:

$$0 \stackrel{!}{=} \delta\Sigma = -\int\limits_{t_1}^{t_2} \int\limits_{\mathbb{R}^3} \Bigg(\pi_\mu \frac{\gamma \mathrm{d}}{\mathrm{d}t} (\delta_x x^\mu) + n\mathrm{q_e} A_\mu \frac{\mathrm{d}}{\mathrm{d}t} (\delta_x x^\mu) + j^\mu \partial_\nu A_\mu \delta_x x^\nu + \\ + j^\mu \delta_A A_\mu + \frac{c}{2Z_0} F^{\mu\nu} \delta_A F_{\mu\nu} + \frac{c}{4Z_0} \delta_x (F^{\mu\nu} F_{\mu\nu}) \Bigg) \mathrm{d}^3 x \mathrm{d}t$$

These six summands can be rewritten as follows:

- The first term can be integrated partially:

$$\delta\Sigma_1 = -\int_{t_1}^{t_2}\int_{\mathbb{R}^3} \pi_\mu \frac{\gamma \mathrm{d}}{\mathrm{d}t}\left(\delta x^\mu\right) \mathrm{d}^3 x \mathrm{d}t =$$

$$= -\left(\int_{\mathbb{R}^3} \gamma \pi_\mu \delta x^\mu \mathrm{d}^3 x\right)_{t_1}^{t_2} + \int_{t_1}^{t_2}\int_{\mathbb{R}^3} \frac{\gamma \mathrm{d}\pi_\mu}{\mathrm{d}t} \delta x^\mu \mathrm{d}^3 x \mathrm{d}t$$

  Since the parameters have to be kept constant at the integration boundaries (at t_1 and t_2), its first summand is zero.

- The same applies to the second term where the charge density varies with the coordinates, but its integral over $\mathbb{R}^3$ is the sum of all charges, which is independent of δx^μ. The time-derivative of the fields can be written explicitly:

$$\delta\Sigma_2 = \int_{t_1}^{t_2}\int_{\mathbb{R}^3} n q_\mathrm{e} \frac{\mathrm{d}A_\mu}{\mathrm{d}t} \delta x^\mu \mathrm{d}^3 x \mathrm{d}t = \int_{t_1}^{t_2}\int_{\mathbb{R}^3} j^\nu \partial_\nu A_\mu \delta x^\mu \mathrm{d}^3 x \mathrm{d}t$$

- In the third and fourth terms one can exchange the summation indices:

$$\delta\Sigma_3 + \delta\Sigma_4 = -\int_{t_1}^{t_2}\int_{\mathbb{R}^3} j^\nu \partial_\mu A_\nu \delta x^\mu \mathrm{d}^3 x \mathrm{d}t - \int_{t_1}^{t_2}\int_{\mathbb{R}^3} j^\nu \delta A_\nu \mathrm{d}^3 x \mathrm{d}t$$

- In the fifth term the potentials A^{μ} are explicitly written, then the integrand is integrated partially over space-time (∂_{μ}):

$$\delta\Sigma_5 = -\int_{t_1}^{t_2}\int_{\mathbb{R}^3} \frac{c}{2Z_0} F^{\mu\nu}\delta F_{\mu\nu}\mathrm{d}^3x\mathrm{dt} = -\int_{t_1}^{t_2}\int_{\mathbb{R}^3} \frac{c}{Z_0} F^{\mu\nu}\partial_{\mu}\delta A_{\nu}\mathrm{d}^3x\mathrm{dt}$$
$$= -\int_{t_1}^{t_2}\int_{\mathbb{R}^3} \frac{c}{Z_0}\partial_{\mu}\left(F^{\mu\nu}\delta A_{\nu}\right)\mathrm{d}^3x\mathrm{dt} + \int_{t_1}^{t_2}\int_{\mathbb{R}^3} \frac{c}{Z_0}\partial_{\mu}F^{\mu\nu}\delta A_{\nu}\mathrm{d}^3x\mathrm{dt}$$

The first, integrated part is equal to

$$-\frac{1}{Z_0}\left(\int_{\mathbb{R}^3} F^{0\nu}\delta A_{\nu}\mathrm{d}^3x\right)\Bigg|_{t_1}^{t_2} - \int_{t_1}^{t_2}\frac{c}{Z_0}\int_{\mathbb{R}^3}\frac{\mathrm{d}}{\mathrm{d}x^i}\left(F^{i\nu}\delta A_{\nu}\right)\mathrm{d}^3x\mathrm{d}t$$

which is zero since the variation is zero at the boundaries t_1, t_2 (first term) and since the fields are assumed to be zero at infinity (second term). Therefore:

$$\delta\Sigma_5 = \int_{t_1}^{t_2}\int_{\mathbb{R}^3} \frac{c}{Z_0}\partial_{\mu}F^{\mu\nu}\delta A_{\nu}\mathrm{d}^3x\mathrm{dt}$$

- In the sixth term the variation can be written explicitly. Similarly to Σ_5, since the fields are zero at infinity and since the coordinates at the integration boundaries are not varied, this term equals zero:

$$\delta\Sigma_6 = -\int_{t_1}^{t_2}\int_{\mathbb{R}^3} \frac{c}{4Z_0}\partial_{\sigma}(F^{\mu\nu}F_{\mu\nu})\delta x^{\sigma}\mathrm{d}^3x\mathrm{dt} = 0$$

Thus, the stationarity condition for the action – with the terms sorted by parameters to be varied – reads:

$$\delta\Sigma = \int_{t_1}^{t_2}\int_{\mathbb{R}^3}\left(\frac{\gamma \mathrm{d}\pi_\mu}{\mathrm{d}t} - j^\nu\left(\partial_\mu A_\nu - \partial_\nu A_\mu\right)\right)\delta x^\mu \mathrm{d}^3x\mathrm{d}t - \\ - \int_{t_1}^{t_2}\int_{\mathbb{R}^3}\left(j^\nu - \frac{c}{Z_0}\partial_\mu F^{\mu\nu}\right)\delta A_\nu \mathrm{d}^3x\mathrm{d}t$$

Since this expression needs to be valid for all times and for all variations of the free parameters, both integrands need to be equal zero independently. Therefore, one obtains the following equations using some of the relations stated in Section A.2:

$$\partial_\mu F^{\mu\nu} = Z_0 j^\nu \tag{A.4}$$

$$\frac{\mathrm{d}\pi_\mu}{\mathrm{d}\tau} = F_{\mu\nu} j^\nu \tag{A.5}$$

Equation A.4 are the inhomogeneous Maxwell's equations (source terms of the fields), and Eq. A.5 is the Lorentz force (fields acting on charges), as will be shown in the following Sections. Both equations are written with covariant expressions only. Solving these two equations iteratively is called "ballistic approach".

With $\mu_0 = 4\pi \cdot 10^{-7}$ kg m/C^2, $\mu_0\varepsilon_0 c^2 \equiv 1$, $Z_0 = \sqrt{\frac{\mu_0}{\varepsilon_0}} = \frac{4\pi\alpha_0\hbar}{\mathrm{e}^2}$ and using the relations in Section A.2 one can confirm the inhomogeneous Maxwell's equations (Gauss's and Ampère-Maxwell laws) [LL97, Jac99]. Under Lorenz gauge, Eq. A.4 can also be written as:

$$\Box A^\mu = Z_0 j^\mu \tag{A.6}$$

As one can easily check, any antisymmetric tensor fulfills:

$$0 = \partial_\mu F_{\nu\sigma} + \partial_\nu F_{\sigma\mu} + \partial_\sigma F_{\mu\nu} \tag{A.7}$$

Thus, the homogeneous Maxwell's equations (Faraday law and Gauss's law for magnetism) reduce to a general property of spacetime:

$$\begin{aligned}\vec{0} =& c\vec{e_i}\frac{\epsilon^{ijk}}{2}\left(\partial_0 F_{jk} + \partial_j F_{k0} + \partial_k F_{0j}\right)\\ =& \vec{e_i}\partial_t\left(\frac{\epsilon^{ijk}}{2}F_{jk}\right) + \vec{e_i}\frac{\epsilon^{ijk}}{2}\left(-\partial_j E_k + \partial_k E_j\right)\\ =& \partial_t\vec{B} - \nabla\times\vec{E}\end{aligned}$$

$$0 = \partial_1 F_{23} + \partial_2 F_{31} + \partial_3 F_{12} = \nabla\cdot\vec{B}$$

For an axisymmetric field and for $r\partial_r B_z \ll B_z$, this leads to the near-axis approximation:

$$B_r \approx -\frac{r}{2}\partial_z B_z \tag{A.8}$$

A.4 Voltage Depression

See [Tsi93, DK81, GC84, CB93] and references therein for details and more sophisticated analysis. Gauss's law in integral form reads

$$\oint_{\partial V} \vec{E} \cdot \mathrm{d}\vec{S} = \frac{Q_{\mathrm{encl}}}{\varepsilon_0}$$

where Q_{encl} is the electric charge enclosed by a volume V with boundary ∂V. For an axisymmetric problem and for a constant charge distribution with $\kappa := \frac{\mathrm{d}Q}{\mathrm{d}z}$, the electric field reduces to

$$E_r = \frac{\kappa(r)}{2\pi\varepsilon_0 r}$$
$$E_\varphi = 0 \qquad E_z = 0$$

The voltage between two radial positions r_1, r_2 is then

$$U(r_1, r_2) = \frac{1}{2\pi\varepsilon_0} \int_{r_1}^{r_2} \frac{\kappa(r)}{r} \mathrm{d}r$$

For a thin beam of charged particles – or, equivalently, for a charged metallic surface – with $\kappa(r) = \kappa_X \cdot \delta(r - r_X)$, the potential reads then

$$\Phi_X(r) = \begin{cases} 0 & , \quad r \leq r_X \\ \frac{\kappa_X}{2\pi\varepsilon_0} \cdot \ln\left(\frac{r}{r_X}\right) & , \quad r_X \leq r \end{cases}$$

For a thin beam with radius r_b in a metallic cylinder with radius R_O, the resulting potential is just the sum of both individual potentials. If the metallic cylinder is grounded, it will charge in such a way that the potential outside of the system (for all $r > R_\mathrm{O}$) equals zero. This can only be achieved if $\kappa_\mathrm{O} = -\kappa_\mathrm{b}$ and if a constant potential $\Phi_\mathrm{d} = \frac{\kappa_\mathrm{b}}{2\pi\varepsilon_0} \cdot \ln\left(\frac{r_\mathrm{b}}{R_\mathrm{O}}\right)$ (i.e. the difference between a "charge-free world" and the ground potential) is added:

$$\Phi(r) = \begin{cases} \frac{\kappa_b}{2\pi\varepsilon_0} \ln\left(\frac{r_b}{R_O}\right) & , \quad r \le r_b \\ \frac{\kappa_b}{2\pi\varepsilon_0} \ln\left(\frac{r_b}{R_O}\right) + \frac{\kappa_b}{2\pi\varepsilon_0} \ln\left(\frac{r}{r_b}\right) & , \quad r_b \le r \le R_O \\ 0 & , \quad R_O \le r \end{cases}$$

The potential Φ_d acts on the electron beam opposite to the voltage accelerating the beam, and is thus termed "voltage depression" (potential drop).

In practice, with κ_b being proportional to the beam current, there is a limitation for the current at which the voltage depression is as large as the accelerating voltage – the "limiting current" –, and a hollow-cavity gyrotron with these parameters is theoretically expected to fail operation.

In the case of a double cylinder (with inner radius R_I, outer radius R_O) with an aligned hollow beam inbetween, the resulting potential is the sum of all three potentials. If the inner and outer cylinders are grounded, which is the case for coaxial insert and cavity wall of a gyrotron, the resulting potential has to be zero for both $r \le R_I$ and $r \ge R_O$:

$$\Phi(r) =$$

$$\begin{cases} 0 & , \quad r \le R_I \\ \frac{\kappa_I}{2\pi\varepsilon_0} \ln\left(\frac{r}{R_I}\right) & , \quad R_I \le r \le r_b \\ \frac{\kappa_I}{2\pi\varepsilon_0} \ln\left(\frac{r}{R_I}\right) + \frac{\kappa_b}{2\pi\varepsilon_0} \ln\left(\frac{r}{r_b}\right) & , \quad r_b \le r \le R_O \\ \frac{\kappa_I}{2\pi\varepsilon_0} \ln\left(\frac{r}{R_I}\right) + \frac{\kappa_b}{2\pi\varepsilon_0} \ln\left(\frac{r}{r_b}\right) + \frac{\kappa_O}{2\pi\varepsilon_0} \ln\left(\frac{r}{R_O}\right) & , \quad R_O \le r \end{cases}$$

One can easily check that $\Phi(r) = 0$ for all $r \ge R_O$ is only fulfilled if

$$\frac{\kappa_I}{\kappa_b} + \frac{\kappa_O}{\kappa_b} = -1$$

$$\frac{\kappa_I}{\kappa_b} = \frac{\ln(r_b/R_O)}{\ln(R_O/R_I)}$$

This leads to a voltage depression (potential at the beam position) of:

$$\Phi_{\mathrm{d}} = \Phi(r_{\mathrm{b}}) = \frac{\kappa_{\mathrm{b}}}{2\pi\varepsilon_0} \frac{\ln(r_{\mathrm{b}}/R_{\mathrm{O}}) \ln(r_{\mathrm{b}}/R_{\mathrm{I}})}{\ln(R_{\mathrm{O}}/R_{\mathrm{I}})} \tag{A.9}$$

For typical gyrotrons, the voltage depression with coaxial insert is by a factor of 5 to 7 smaller than for the same configuration without coaxial insert, mostly due to the small ratio $r_{\mathrm{b}} : R_{\mathrm{I}}$. In the limit $R_{\mathrm{I}} \to 0$, the terms including R_{I} in Eq. A.9 cancel each other and the result is the same as in the hollow-cavity case.

It should be noted that in a realistic (thick) electron beam in a gyrotron cavity with axially dependent radii, the details of the above considerations are a bit more complicated, but the general results are very much the same. In a realistic beam, voltage depression is different for different radial positions of the electrons and thus contributes to the energy spread of the electrons. It can also be dependent of the azimuthal coordinate if beam, outer wall, and inner rod are misaligned versus each other, or can even be dependent of the longitudinal coordinate if the three physical entities are tilted towards each other. Furthermore, electron beams in gyrotrons tend to neutralize (at least partly) during CW operation due to residual atoms/ions in the tube. This effect screens the charge of the beam and reduces the severity of the limiting current in hollow-cavity gyrotrons. The residual voltage depression on a typical partly neutralized beam in a hollow cavity is similar to the depression on a nonneutralized beam in the same cavity with coaxial insert.

Only in some cases, voltage depression can be expressed analytically using mirror charges and existing symmetries. For treatment of more general geometries, see [FAG+16a].

A.5 Plasma Frequency

Using some of the relations given in Section A.2, the Lagrange density (Eq. A.3) reads

$$\Lambda = -\left(\frac{\mathrm{m_e}}{\mathrm{q_e^2}n} j_\mu + A_\mu\right) j^\mu - \frac{c}{4Z_0} F^{\mu\nu} F_{\mu\nu} \tag{A.10}$$

Especially in a plasma (of particles with mass $\mathrm{m_e}$) the canonical momentum (term in brackets) has to be regarded as the relevant quantity. For a stationary plasma, this term should vanish. (In this case, the Hamiltonian remains constant if changes in the electric potential are neglected, i.e. the canonical momentum acts as an adiabatic invariant.) If one now substitutes $A^\mu = -\frac{\mathrm{m_e}}{\mathrm{q_e^2}n} j^\mu$ in the inhomogeneous Maxwell's Equation A.6, one obtains

$$\Box j^\mu = -\frac{Z_0 \mathrm{q_e^2} n}{\mathrm{m_e} c} j^\mu$$

For each space-time component of this current density, this is the simplest form of an inhomogeneous wave equation. One can see that the quantity $\omega_\mathrm{P} = \sqrt{\frac{c Z_0 \mathrm{q_e^2} n}{\mathrm{m_e}}}$ plays the role of a Lorentz-invariant angular frequency: it is not dependent on the wavelength $2\pi/k$. ω_P is called the plasma frequency. In a plasma consisting of several particle species with different masses, such as electrons and deuterium/tritium ions, the effective plasma frequency

$$\omega_\mathrm{P} = \sqrt{\sum_{\text{species } i} {\omega_{P,i}}^2}$$

is close to the electron plasma frequency because of their smallest mass. Nonrelativistic, but more strict derivations of ω_P can be found in e.g. [RS83, Sta05, Sta10].

A.6 Lorentz Force

Since the electromagnetic field is usually induced by a vast number of particles, the source terms in the Maxwell's equations are expressed as charge densities. The Lorentz force, on the other hand, is usually viewed as acting on individual charged (macro-)particles. Spacial integration of Eq. A.5 in one inertial frame (e.g. the lab frame) over a point-like particle with four-momentum p^{μ} and charge $\mathrm{q_e}$ yields:

$$\frac{\mathrm{d}p^{\mu}}{\mathrm{d}t} = \mathrm{q_e} F^{\mu\nu} v_{\nu} \tag{A.11}$$

The quantity $v^{\nu} \equiv y^{\nu}/\gamma$ equals $\begin{pmatrix} c \\ \vec{v} \end{pmatrix}$. If μ runs over the spacial indices and $\nu = 0$, one obtains the expression for the electric force:

$$\frac{\mathrm{d}\vec{p}}{\mathrm{d}t} = \mathrm{q_e} \vec{E} \tag{A.12}$$

If both μ and ν run over the spacial indices, one obtains the expression for the magnetic force:

$$\frac{\mathrm{d}\vec{p}}{\mathrm{d}t} = \mathrm{q_e} \vec{v} \times \vec{B} \tag{A.13}$$

If $\mu = 0$ and ν runs over the spacial indices, one obtains the energy gain of the charged particle (accelerating power) of the interaction:

$$P := \frac{\mathrm{d}W}{\mathrm{d}t} = \mathrm{q_e} \vec{v} \cdot \vec{E} \tag{A.14}$$

While the magnetic field component accelerates the particle (Eq. A.13) it does not change its kinetic energy.

A.7 Motion of a Charged Particle in an Electromagnetic Field

See e.g. the corresponding sections in [LL97] and [Sta05].

A.7.1 General Considerations; Cyclotron Motion

The three-dimensional quantities $X = E, B, p, v$ in the Lorentz force equation (Eq. A.11) can be split into their axial component $X_{\parallel} \equiv X_z \equiv X^3$ and their transverse components $X_x \equiv X^1$, $X_y \equiv X^2$, which themselves can be combined into complex-valued "transverse" variables $X_{\perp} := X^1 + \mathrm{i}X^2$. This leads to one equation for the transverse momentum

$$\frac{\mathrm{d}p_{\perp}}{\mathrm{d}t} = \mathrm{q_e}\left(F^{1\nu} + \mathrm{i}F^{2\nu}\right) v_{\nu} \tag{A.15a}$$

$$= \mathrm{q_e} E_{\perp} + \mathrm{i}\mathrm{q_e} v_{\parallel} B_{\perp} - \mathrm{i}\mathrm{q_e} v_{\perp} B_{\parallel} \tag{A.15b}$$

and two equations for the axial momentum and the transferred power:

$$\frac{\mathrm{d}p_{\parallel}}{\mathrm{d}t} = \mathrm{q_e} E_{\parallel} + \mathrm{q_e}\mathrm{Im}(v_{\perp}^* B_{\perp}) \tag{A.16}$$

$$\frac{\mathrm{d}W}{\mathrm{d}t} = \mathrm{q_e} v_{\parallel} E_{\parallel} + \mathrm{q_e}\mathrm{Re}(v_{\perp}^* E_{\perp}) \tag{A.17}$$

The equations of motion of a charged particle in a constant magnetic field $B_{\parallel}$ are then:

$$\frac{\mathrm{d}p_{\perp}}{\mathrm{d}t} = -\mathrm{i}\mathrm{q_e} v_{\perp} B_{\parallel}$$

$$\frac{\mathrm{d}p_{\parallel}}{\mathrm{d}t} = 0$$

$$\frac{\mathrm{d}W}{\mathrm{d}t} = 0$$

With the ansatz $p_\perp = \hat{p}e^{-\mathrm{i}\Omega t}$ and $p_\perp = \gamma \mathrm{m_e} v_\perp$, one obtains the cyclotron frequency[1] of a particle with mass $\mathrm{m_e}$ and charge $\mathrm{q_e}$:

$$\Omega_\mathrm{C} = \frac{\mathrm{q_e} B_\|}{\gamma \mathrm{m_e}} \tag{A.18}$$

The sign indicates the direction of the movement, which, with the z-axis pointing towards the observer, is clockwise (left-hand screw around the z-axis) for a positively charged particle. One can see that the cyclotron frequency *increases* with *decreasing* particle energy (γ) and vice versa.

The cyclotron radius is then given by

$$r_\mathrm{C} = \frac{c\beta_\perp}{\Omega_\mathrm{C}} = \frac{\beta_\perp \gamma \mathrm{m_e} c}{\mathrm{q_e} B_\|} = \frac{p_\perp}{\mathrm{q_e} B_\|} \tag{A.19}$$

A.7.2 Motion in Crossed Fields

To consider the general case of crossed electric and magnetic fields, one can choose coordinates such that $B_\perp$ vanishes:

$$\begin{aligned}
\frac{\mathrm{d}p_\perp}{\mathrm{d}t} &= \mathrm{q_e} E_\perp - \mathrm{i}\mathrm{q_e} v_\perp B_z \\
\frac{\mathrm{d}p_z}{\mathrm{d}t} &= \mathrm{q_e} E_z \\
\frac{\mathrm{d}W}{\mathrm{d}t} &= \mathrm{q_e} v_z E_z + \mathrm{q_e}\mathrm{Re}(v_\perp^* E_\perp)
\end{aligned}$$

The ansatz $p_\perp = \hat{p}e^{\mathrm{i}\Omega t} + p_\mathrm{dr}$, where p_dr is time-independent, yields

$$\mathrm{i}\Omega \hat{p} e^{\mathrm{i}\Omega t} = \mathrm{q_e} E_\perp - \mathrm{i}\frac{\mathrm{q_e} B_z}{\gamma \mathrm{m_e}} \hat{p} e^{\mathrm{i}\Omega t} - \mathrm{i}\frac{\mathrm{q_e} B_z}{\gamma \mathrm{m_e}} p_\mathrm{dr}$$

[1] This frequency has to be distinguished strictly from the Larmor frequency $\Omega_\mathrm{Larmor} = \frac{\mathrm{q_e} B_\|}{2\mathrm{m_e}}$, which relates the magnetic field acting additionally to a central potential to the Coriolis force in a rotating coordinate system [FLS62].

One knows from the previous considerations that the term on the left hand side is equal to the second term on the right hand side (it is just the cyclotron motion), hence one obtains a charge-independent drift velocity perpendicular to both the electric and the magnetic field:

$$v_{\mathrm{dr}} = -\mathrm{i}\frac{E_\perp}{B_z}$$

The power transferred to the particle simplifies then to

$$\frac{\mathrm{d}W}{\mathrm{d}t} = \mathrm{q_e} v_z E_z + \frac{\mathrm{q_e}\hat{p}}{\gamma \mathrm{m_e}} \mathrm{Re}(e^{-\mathrm{i}\Omega t} E_\perp)$$

where the right term averages out over one cyclotron period.

A.7.3 Grad-B Drift

Another important situation is a weakly inhomogeneous magnetic field $\vec{B} = B_z(x,y)\,\vec{e_z}$. Starting again from

$$\frac{\mathrm{d}p_\perp}{\mathrm{d}t} = -\mathrm{i}\mathrm{q_e} v_\perp B_z$$

one can expand the quantities $p_\perp$, $v_\perp$ and B_z in values for the homogeneous magnetic field at the origo plus small correction terms:

$$\begin{aligned} p_\perp &= p_\perp^{(0)} + p_\perp^{(1)} \\ v_\perp &= v_\perp^{(0)} + v_\perp^{(1)} \\ B_z &= B_z^{(0)} + (x\partial_x + y\partial_y) B_z^{(0)} \end{aligned}$$

Here, the interesting quantity is $v_\perp^{(1)}$ averaged over one cyclotron period, with $\langle X \rangle = \frac{1}{\tau}\int\limits_t^{t+\tau} X(t')\mathrm{d}t'$. The corresponding Lorentz force reads now:

$$\begin{aligned}
&\frac{\mathrm{i}}{\mathrm{q_e}}\frac{\mathrm{d}}{\mathrm{d}t}\left\langle p_\perp^{(0)}\right\rangle + \frac{\mathrm{i}}{\mathrm{q_e}}\frac{\mathrm{d}}{\mathrm{d}t}\left\langle p_\perp^{(1)}\right\rangle = \\
&= \left\langle v_\perp^{(0)}\right\rangle B_z^{(0)} + \left\langle v_\perp^{(0)}(x\partial_x + y\partial_y)\right\rangle B_z^{(0)} + \\
&+ \left\langle v_\perp^{(1)}\right\rangle B_z^{(0)} + \left\langle v_\perp^{(1)}(x\partial_x + y\partial_y)\right\rangle B_z^{(0)}
\end{aligned}$$

The first terms on both sides are just the homogeneous-field case and thus cancel out from the equation. The rightmost term in brackets is the product of two small values and can thus be ignored. One looks for solutions where $\langle v_\perp^{(1)}\rangle = \mathrm{const.}$, thus $\mathrm{d}\langle p_\perp^{(1)}\rangle/\mathrm{d}t \overset{!}{=} 0$. Dropping the (0)'s and writing $v_\perp^{(1)} \equiv v_{\mathrm{dr}}$ leads to:

$$\langle v_{\mathrm{dr}}\rangle B_z = -\left\langle v_\perp(x\partial_x + y\partial_y)\right\rangle B_z$$

Assume that the field is sufficiently homogeneous such that $r_{\mathrm{C}} = \mathrm{const.}$, i.e. $r_{\mathrm{C}} \ll \frac{B_z}{|\nabla B_z|}$. With $x_\perp = r_{\mathrm{C}}(\cos(\Omega t) + \mathrm{i}\sin(\Omega t))$ and consequently $v_\perp = \dot{x}_\perp$, one can average over the products of sines and cosines:

$$\begin{aligned}
\langle v_{\mathrm{dr}}\rangle &= -\frac{\mathrm{i}\Omega r_{\mathrm{C}}^2}{B_z}\big\langle\cos^2(\Omega t)\partial_x + \sin(\Omega t)\cos(\Omega t)\partial_y + \\
&\quad + \mathrm{i}\sin(\Omega t)\cos(\Omega t)\partial_x + \mathrm{i}\sin(\Omega t)^2\partial_y\big\rangle B_z \\
&= -\frac{\mathrm{i}\Omega r_{\mathrm{C}}^2}{B_z}\left\langle\cos^2(\Omega t)\partial_x + \mathrm{i}\sin(\Omega t)^2\partial_y\right\rangle B_z \\
&= -\frac{\mathrm{i}\Omega r_{\mathrm{C}}^2}{2B_z}\partial_\perp B_z \equiv -\frac{\mathrm{i}\mathrm{q_e} r_{\mathrm{C}}^2}{2\gamma\mathrm{m_e}}\partial_\perp B_z \equiv -\frac{\mathrm{i}W_{\mathrm{C}}}{\mathrm{q_e}B_z^2}\partial_\perp B_z
\end{aligned}$$

As a conclusion, an inhomogeneous magnetic field causes an average drift of charged particles perpendicular to the field direction and to the field gradient. Consequently, this drift is termed Grad-B drift. The drift depends on the particle charge, i.e. it separates positive and negative charges.

A.8 Magnetic Compression and Busch's Theorem

The magnetic flux Ψ through a surface S_0 is defined as

$$\Psi_0 := \int_{S_0} \vec{B} \cdot \mathrm{d}\vec{S}$$

where $\mathrm{d}\vec{S}$ is a directed surface element. Application of Green's theorem on $\nabla \cdot \vec{B} = 0$ with $\partial V = \sum_i S_i$ results in

$$0 = \oint_{\partial V} \vec{B} \cdot \mathrm{d}\vec{S} \equiv \sum_i \Psi_i$$

If the total surface consists of three parts with $\vec{B} \parallel \mathrm{d}\vec{S}$ for all surface elements $\mathrm{d}\vec{S}$ of surfaces with indices $i = 1, 2$ and $\vec{B} \perp \mathrm{d}\vec{S}$ for $i = 3$, it follows that $B_1 S_1 = B_2 S_2$. For the special case of circles $S_i = \pi r_i^2$, which is relevant for axisymmetric magnetic fields:

$$B_1 r_1^2 = B_2 r_2^2$$

For a hollow beam of electrons (which follow the magnetic field lines, see Section A.7.1), these radii describe the beam radius at the respective magnetic fields. If such a beam enters an increasing magnetic field, it is *magnetically compressed.*

For particles in electromagnetic fields, conservation laws apply only for canonical, but not for kinematic observables.[2] Consider now a single particle in a magnetic field. Since no energy is transferred to or from the particle, the canonical momentum of this system will remain constant [Bur01]. Integration over a closed loop (around surface S) yields the *adiabatic invariant*

[2] Obviously, the kinetic momentum is not conserved for an electron accelerated in a static electric field, just to mention a simple example.

$$\text{const.} = \oint_{\partial S} \left(\vec{p} + \mathrm{q_e}\vec{A}\right) \cdot \mathrm{d}\vec{l}$$

The second part of the integral can be written (using Stoke's theorem) as a surface integral $\mathrm{q_e} \int_S (\nabla \times \vec{A}) \cdot \mathrm{d}\vec{S}$, which is equal to $\mathrm{q_e}\Psi$ because of $\vec{B} = \nabla \times \vec{A}$. In the case of a homogeneous magnetic field, $\oint_{\partial S} \vec{p} \cdot \mathrm{d}\vec{l} = 2\pi r p$ and $r = r_\mathrm{C} = \text{const.}$, thus

$$\text{const.} = \mathrm{q_e}\Psi + 2\pi r_\mathrm{C} p \tag{A.20}$$

With $\Psi = \pi r_\mathrm{C}^2 B$, one obtains for changes in the perpendicular momentum:

$$\Delta p = -\frac{\mathrm{q_e} r_\mathrm{C}}{2} \Delta B$$

In electron beam optics, this result is termed *Busch's theorem*, after [Bus26]. Another popular formulation of this theorem is using $p = \mathrm{m_e} r_\mathrm{C} \dot{\varphi}$ and assuming the initial momentum p_st to be zero:

$$\dot{\varphi} = -\frac{\mathrm{q_e}}{2\pi \mathrm{m_e} r_\mathrm{C}^2} (\Psi - \Psi_\mathrm{st})$$

A.9 Resonance Condition

With respect to a chosen axis, the vector $\vec{k}$ of a plane electromagnetic wave consists of an axial component k_z and a transversal component $k_\perp$ with $k^2 = k_\perp^2 + k_z^2$. Standing waves have $k_z = 0$ and are termed "at cutoff"; their frequency is $\omega = \omega_\perp = ck_\perp$. If one moves in a frame with velocity v_{gr} relative to a standing wave, its frequency is increased by the relativistic factor between the two frames, i.e.

$$\omega = \frac{\omega_\perp}{\sqrt{1 - \left(\frac{v_{\mathrm{gr}}}{c}\right)^2}}$$

This expression is equivalent to the definition of the group velocity $v_{\mathrm{gr}} := \frac{\mathrm{d}\omega}{\mathrm{d}k_z}$ with ω following from the wave vector decomposition as:

$$\omega(k_z) = \sqrt{\omega_\perp^2 + c^2 k_z^2} \tag{A.21}$$

Let now a charged particle gyrate in a weak plane wave, having no p_z. Within a time interval $[0;\tau]$ or within a distance $L := v_{\mathrm{gr}}\tau$, the gyrating particle remains in the same period with a weak rotating electric field $E_\perp = \hat{E}e^{\mathrm{i}(\omega t - k_z z)}$ as long as their phases do not differ by more than 2π. This resonance condition can also be fulfilled for higher harmonics of the cyclotron frequency $s\,\Omega_{\mathrm{C}}$:

$$\begin{aligned} \omega\tau - k_z L \leq s\,\Omega_{\mathrm{C}}\,\tau \leq \omega\tau - k_z L + 2\pi \\ \Leftrightarrow \omega - s\,\Omega_{\mathrm{C}} \leq k_z\,v_{\mathrm{gr}} \leq \omega - s\Omega_{\mathrm{C}} + \frac{2\pi}{\tau} \end{aligned}$$

Exact resonance would be given for $\tau \to \infty$, provided that little or no energy is exchanged between particle and wave (orthogonality: $\vec{v} \cdot \vec{E} \approx 0$):

$$\omega(k_z) \approx s\,\Omega_{\mathrm{C}} + k_z v_{\mathrm{gr}} \tag{A.22}$$

The term $k_z v_{\mathrm{gr}}$ is known as Doppler term; it is clear that the resonance condition is independent of the chosen reference frame, one could for example go back to the lab frame where the wave is at cutoff and the electron moves at v_{gr}. The interaction between electron beam and microwave field in a cavity can thus be universally illustrated in a Brillouin diagram (dispersion diagram; see e.g. [Thu16]), where the frequency ω of the waveguide mode is displayed as a function of the z-component of the wavevector, k_z: Equation A.21 describes the dispersion line, a hyperbola, while Eq. A.22 is the straight beam line. Possible interaction points are the intersections of line and hyperbola, which can be either one point (in the case of gyrotrons) or two points, of which one typically lies in the forward- ($k_z > 0$) and one in the backward-wave region ($k_z < 0$).

In LOGs, azimuthal index and harmonics number of the mode must agree, $m \stackrel{!}{=} s$, for optimum operation.

A.10 Diffractive Quality Factor

Diffractive quality factor and output power are per definition related with frequency and energy content by

$$Q_{\mathrm{diff}} P_{\mathrm{out}} = \omega W \tag{A.23}$$

The radiated power in axial direction P_{out} is just the stored energy moving at group velocity along one cavity length, $\frac{W v_{\mathrm{gr}}}{L_{\mathrm{O}}}$, whereas the group velocity can be expressed as $v_{\mathrm{gr}} = \frac{c^2 k_z}{\omega}$ (see Eq. A.21). For a wave with axial index $l = 1$ one can assume $k_z \lesssim \frac{\pi}{L_{\mathrm{O}}}$, hence

$$Q_{\mathrm{diff}} \gtrsim 4\pi \left(\frac{L_{\mathrm{O}}}{\lambda} \right)^2 \tag{A.24}$$

B Theory of TE Modes in Coaxial Gyrotron Cavities

B.1 Transversal Field Profile

Maxwell's equations written in complex transverse coordinates (where, if translational symmetry is assumed, $X_{\|} \equiv X_z$ at any position z for the fields $X = E, B$) read as follows:

$$\frac{1}{r}\partial_{\perp}^{*} r E_{\perp} = -\mathrm{i}\partial_t B_{\|} - \partial_{\|} E_{\|} + \frac{\rho}{\varepsilon_0} \tag{B.1a}$$

$$-\mathrm{i}\partial_{\perp} E_{\|} + \mathrm{i}\partial_{\|} E_{\perp} = -\partial_t B_{\perp} \tag{B.1b}$$

$$-\mathrm{i}\partial_{\perp} B_{\|} + \mathrm{i}\partial_{\|} B_{\perp} = \frac{1}{c^2}\partial_t E_{\perp} + \mu_0 j_{\perp} \tag{B.1c}$$

$$\frac{1}{r}\partial_{\perp}^{*} r B_{\perp} = \frac{\mathrm{i}}{c^2}\partial_t E_{\|} - \partial_{\|} B_{\|} + \mu_0 j_{\|} \tag{B.1d}$$

With absence of sources, the harmonic ansatz $X = \hat{X}(r)e^{\mathrm{i}\omega t}e^{-\mathrm{i}k_{\|}z}e^{-\mathrm{i}m\varphi}$ for the field components and considering only TE waves ($E_{\|} = 0$), one obtains $\partial_{\perp} = \partial_r - \frac{m}{r}$, $\partial_{\perp}^{*} = \partial_r + \frac{m}{r}$ and

$$\frac{1}{r}\partial_{\perp}^{*} r \hat{E}_{\perp} = \omega \hat{B}_{\|} \tag{B.2a}$$

$$k_{\|}\hat{E}_{\perp} = -\mathrm{i}\omega \hat{B}_{\perp} \tag{B.2b}$$

$$-\mathrm{i}\partial_{\perp}\hat{B}_{\|} + k_{\|}\hat{B}_{\perp} = \frac{\mathrm{i}\omega}{c^2}\hat{E}_{\perp} \tag{B.2c}$$

$$\frac{1}{r}\partial_{\perp}^{*} r \hat{B}_{\perp} = -\partial_{\|} B_{\|} \tag{B.2d}$$

Applying the operator $r\partial_\perp^* r$ on Eqs. B.2b and B.2c yields

$$\partial_\perp^* r\hat{E}_\perp = \omega r\hat{B}_\parallel \tag{B.3a}$$

$$\partial_\perp^* r\hat{B}_\perp = \frac{\mathrm{i}k_\parallel}{\omega}\partial_\perp^* r\hat{E}_\perp = \mathrm{i}k_\parallel r\hat{B}_\parallel \tag{B.3b}$$

$$r\partial_\perp^* r\partial_\perp\hat{B}_\parallel + \mathrm{i}rk_\parallel\partial_\perp^* r\hat{B}_\perp + r\frac{\omega}{c^2}\partial_\perp^* r\hat{E}_\perp = 0 \tag{B.3c}$$

Substituting Eqs. B.3a and B.3b into Eq. B.3c and using $\omega \equiv ck$ and $k_\perp^2 = k^2 - k_\parallel^2$, one obtains

$$r\partial_\perp^* r\partial_\perp\hat{B}_\parallel + r^2 k_\perp^2\hat{B}_\parallel = 0 \tag{B.4}$$

Executing the transverse derivative and with $\frac{1}{k_\perp}\partial_r \equiv \frac{\mathrm{d}}{\mathrm{d}(k_\perp r)}$, Eq. B.4 can now be written as the Bessel differential equation

$$k_\perp^2 r^2\frac{\mathrm{d}^2\hat{B}_\parallel}{\mathrm{d}(k_\perp r)^2} + k_\perp r\frac{\mathrm{d}\hat{B}_\parallel}{\mathrm{d}(k_\perp r)} + \left(k_\perp^2 r^2 - m^2\right)\hat{B}_\parallel = 0 \tag{B.5}$$

Thus, $\hat{B}_\parallel$ can be expressed as a linear combination of Bessel and Neumann functions

$$\hat{B}_{\parallel(,m)}(r) = \hat{B}_J \cdot J_m(k_\perp r) + \hat{B}_N \cdot N_m(k_\perp r) \tag{B.6}$$

with constant coefficients $\hat{B}_J$ and $\hat{B}_N$ which depend on the imposed boundary conditions, e.g. at two radii R_I, R_O. For a derivation using standard cylindrical coordinates, see e.g. [Mei66] Volume 2, II.4, eq.s 234ff. With eigenvalue $\chi_\mathrm{O} := k_\perp R_\mathrm{O}$ resulting from a solution respecting the imposed boundary conditions and with the (absolute) caustic radius $|m| =: k_\perp R_\mathrm{K}$ of a mode, notice how the (normalized) area occupied by a solution, $(\chi^2 - m^2)$, appears naturally in the Bessel equation. For each m, several roots exist, numbered with $p = 1, 2, ...$; hence, $\chi_\mathrm{O} \equiv \chi_{mp}$.

It should be emphasized here that the azimuthal index m can be both a positive or a negative integer (or zero). The negative sign before m in the harmonic ansatz ensures that if m is positive, the mode rotates in mathematically positive direction (i.e., in the same direction as the electrons rotate if $B_{\|} > 0$). This is the same convention as in e.g. [Avr06], and opposite to [Ker96]. In order to remove ambiguity, the notation $\mathrm{TE}^{\mathrm{ct}}_{mp}$ for $\mathrm{TE}_{-|m|,p}$ has been introduced by K. Avramidis [Avr06] and is used in this work for the same purpose. Since $J_{-m} = (-1)^m J_m$ (see [AS64], eq. 9.1.5) and since in most formulas only m^2 appears, the sign does not play a significant role and will be omitted in formulas where it is obvious that only nonnegative m are reasonable. The most prominent exception is the definition for the components of $\hat{E}_{\perp}$:

$$\hat{E}_{\varphi} = \mathrm{i}\frac{\omega}{k_{\perp}^2}\partial_r \hat{B}_{\|} \tag{B.7a}$$

$$\hat{E}_r = -\frac{m}{r}\frac{\omega}{k_{\perp}^2}\hat{B}_{\|} \tag{B.7b}$$

Using these explicit expressions, one can use a special property of the Bessel (and Neumann) functions to write $\hat{E}_{\perp}$ as follows:

$$\hat{E}_{\perp} = \hat{E}_r + \mathrm{i}\hat{E}_{\varphi} = -\frac{\omega}{k_{\perp}}\left(\frac{m}{k_{\perp}r}\hat{B}_{\|,m} + \frac{\mathrm{d}\hat{B}_{\|,m}}{\mathrm{d}(k_{\perp}r)}\right) = -\frac{\omega}{k_{\perp}}\hat{B}_{\|,m-1} \tag{B.8}$$

$$\hat{E}_{\perp}^{*} = -\frac{\omega}{k_{\perp}}\hat{B}_{\|,m+1} \tag{B.9}$$

$$|E_{\perp}| = \frac{\omega}{k_{\perp}}\sqrt{\left(\hat{B}_{\|,m-1}^2 + \hat{B}_{\|,m+1}^2\right)/2} \tag{B.10}$$

A rather general boundary condition (to determine $k_{\perp}$ and to fix one of the free parameters $\hat{B}_J$, $\hat{B}_N$) is now an impedance boundary condition at two radii in the cylinder-symmetric system: inner radius R_{I} and outer radius R_{O} (see e.g. [Ker96]). Such a boundary condition can be imposed by a metallic wall with a surface structure that behaves like an array of small resonant circuits (see e.g. the considerations on split-ring resonators as metamaterials [BBM+05, Shv14]); it includes the usual Neumann boundary condition as a special case. Since the electric component of the wave corresponds to the voltage between two surface structure elements and since $E_{\|} = 0$, the E_φ component is the only useful. Coupling E_φ and B_r (e.g. by small circuits parallel to the wall surface) would result in a real-valued resistance; thus, only a coupling to B_z is feasible. This leads to the two following boundary conditions on the inner/outer surface:

$$\begin{aligned} Z_{\mathrm{I/O}} :=& \frac{\mu_0 \hat{E}_\varphi(R_{\mathrm{I/O}})}{\hat{B}_{\|}(R_{\mathrm{I/O}})} \equiv \frac{\mathrm{i}\mu_0\omega}{k_\perp^2} \frac{\partial_r \hat{B}_{\|}(R_{\mathrm{I/O}})}{\hat{B}_{\|}(R_{\mathrm{I/O}})} \\ \equiv& \frac{\mathrm{i}\mu_0\omega}{k_\perp} \cdot \frac{\hat{B}_J J_m'(k_\perp R_{\mathrm{I/O}}) + \hat{B}_N N_m'(k_\perp R_{\mathrm{I/O}})}{\hat{B}_J J_m(k_\perp R_{\mathrm{I/O}}) + \hat{B}_N N_m(k_\perp R_{\mathrm{I/O}})} \end{aligned} \tag{B.11}$$

One can easily see that the unknown coefficients $\hat{B}_J$ and $\hat{B}_N$ are each identical in these two equations if the characteristic equation (with $k_\perp R_{\mathrm{I/O}} = \chi_{\mathrm{I/O}}$ and $C_R = R_{\mathrm{O}}/R_{\mathrm{I}}$) is fulfilled:

$$\begin{aligned} &\left(\frac{k_\perp Z_{\mathrm{O}}}{\mathrm{i}\mu_0\omega} J_m(\chi_{\mathrm{O}}) - J_m'(\chi_{\mathrm{O}})\right)\left(\frac{k_\perp Z_{\mathrm{I}}}{\mathrm{i}\mu_0\omega} N_m\left(\frac{\chi_{\mathrm{O}}}{C_R}\right) - N_m'\left(\frac{\chi_{\mathrm{O}}}{C_R}\right)\right) = \\ &= \left(\frac{k_\perp Z_{\mathrm{I}}}{\mathrm{i}\mu_0\omega} J_m\left(\frac{\chi_{\mathrm{O}}}{C_R}\right) - J_m'\left(\frac{\chi_{\mathrm{O}}}{C_R}\right)\right)\left(\frac{k_\perp Z_{\mathrm{O}}}{\mathrm{i}\mu_0\omega} N_m(\chi_{\mathrm{O}}) - N_m'(\chi_{\mathrm{O}})\right) \end{aligned} \tag{B.12}$$

(Mind that [AS64], 9.1.34 does not simplify this relation.) From this equation, one can obtain the famous eigenvalue curves $\chi_{\mathrm{O}}(C_R)$.

It is also apparent that, if Z_I and Z_O are known, one of the two values $\hat{B}_J$, $\hat{B}_N$ is determined as well. One can, for example, conveniently define the field amplitude

$$\begin{aligned}\hat{B}_m :=& \frac{\hat{B}_{J,m}}{\frac{k_\perp Z_\mathrm{I}}{\mathrm{i}\mu_0\omega} N_m(k_\perp R_\mathrm{I}) - N'_m(k_\perp R_\mathrm{I})} \\ \equiv& \frac{-\hat{B}_{N,m}}{\frac{k_\perp Z_\mathrm{I}}{\mathrm{i}\mu_0\omega} J_m(k_\perp R_\mathrm{I}) - J'_m(k_\perp R_\mathrm{I})}\end{aligned} \tag{B.13}$$

which can be used to rewrite Eq. B.6 as

$$\begin{aligned}\hat{B}_{\|,m}(r) =& \hat{B}_m \cdot \Big(J_m\left(k_\perp r\right), -N_m\left(k_\perp r\right) \Big) \cdot \\ & \cdot \begin{pmatrix} N'_m\left(k_\perp R_\mathrm{I}\right) & N_m\left(k_\perp R_\mathrm{I}\right) \\ J'_m\left(k_\perp R_\mathrm{I}\right) & J_m\left(k_\perp R_\mathrm{I}\right) \end{pmatrix} \cdot \begin{pmatrix} -1 \\ \frac{k_\perp Z_\mathrm{I}}{\mathrm{i}\mu_0\omega} \end{pmatrix}\end{aligned} \tag{B.14}$$

Concerning th characteristic equation, Eq. B.12, there are some special cases which one might consider:

1. If the radii ratio C_R goes to infinity (i.e. if the inner radius goes to zero), all values of Neumann functions with small arguments become much larger than the values of Bessel functions; thus,

$$\frac{k_\perp Z_\mathrm{O}}{\mathrm{i}\mu_0\omega} J_m\left(\chi_\mathrm{O}\right) - J'_m\left(\chi_\mathrm{O}\right) = 0$$

2. If the impedance at the outer wall Z_O is zero, the equation can be simplified to

$$\begin{aligned} J'_m\left(\chi_\mathrm{O}\right) \left(\frac{k_\perp Z_\mathrm{I}}{\mathrm{i}\mu_0\omega} N_m\left(\frac{\chi_\mathrm{O}}{C_R}\right) - N'_m\left(\frac{\chi_\mathrm{O}}{C_R}\right)\right) =& \\ = N'_m\left(\chi_\mathrm{O}\right) \left(\frac{k_\perp Z_\mathrm{I}}{\mathrm{i}\mu_0\omega} J_m\left(\frac{\chi_\mathrm{O}}{C_R}\right) - J'_m\left(\frac{\chi_\mathrm{O}}{C_R}\right)\right)&\end{aligned} \tag{B.15}$$

For $Z_{\mathrm{O}} = 0$, it follows from Eq. B.11 that

$$0 = \hat{B}_J J'_m(\chi_{\mathrm{O}}) + \hat{B}_N N'_m(\chi_{\mathrm{O}})$$

If the radii ratio C_R goes to infinity,

$$J'_m(\chi_{\mathrm{O}}) = 0$$

One can see that the impedance on the inner surface does not play a role anymore, as it should be.

3. If the impedance at the inner wall Z_{I} is zero, the equation simplifies to

$$N'_m\left(\frac{\chi_{\mathrm{O}}}{C_R}\right)\left(\frac{k_\perp Z_{\mathrm{O}}}{\mathrm{i}\mu_0\omega} J_m(\chi_{\mathrm{O}}) - J'_m(\chi_{\mathrm{O}})\right) = \\ = J'_m\left(\frac{\chi_{\mathrm{O}}}{C_R}\right)\left(\frac{k_\perp Z_{\mathrm{O}}}{\mathrm{i}\mu_0\omega} N_m(\chi_{\mathrm{O}}) - N'_m(\chi_{\mathrm{O}})\right)$$

This is formally the same equation as above if one exchanges χ_{O} with χ_{I} and Z_{I} with Z_{O}.

4. If the impedance at both walls is zero (or if the frequency ω goes to infinity), the equation simplifies to the cross-product (see [AS64] 9.5.30/31)

$$J'_m(\chi_{\mathrm{O}}) \cdot N'_m\left(\frac{\chi_{\mathrm{O}}}{C_R}\right) = N'_m(\chi_{\mathrm{O}}) \cdot J'_m\left(\frac{\chi_{\mathrm{O}}}{C_R}\right)$$

As above, if the radii ratio C_R goes to infinity,

$$J'_m(\chi_{\mathrm{O}}) = 0$$

Of course, one would arrive at the same condition if one just takes the singularity-free solution of Eq. B.5, $\hat{B}_{\parallel}(r) = \hat{B} J_m(k_\perp r)$, and respects the boundary condition of a perfectly conducting surface $E_\varphi(R_\mathrm{O}) = 0$ (or equivalently $B_r(R_\mathrm{O}) = 0$). The canonical numbering of eigenvalues is such that $\chi_\mathrm{O} \equiv \chi_{mp}$ is the p-th root of $J'_m(\chi_{mp}) = 0$.

It is common to characterize TE modes by their (relative) caustic radius $C_\mathrm{K} = |m|/\chi_{mp}$. Modes with large caustic radius are termed "whispering-gallery modes", while those with small caustic are termed "volume modes". While this distinction has originally been made for hollow-cavity modes, it also applies to coaxial-cavity modes. Due to the dependence of χ_O on p the relation between χ_O and m can readily be translated to a relation between m and p. In the frame of this work it appeared useful to automatically scan a certain area in the m-p plane for modes of either type. For these purposes, a variable termed "mode character" h has been introduced, which can take real values around the interval from 0 to 1 and returns the following p for given m:

$$p_{min} = \max\{0.6 \cdot h - 0.1; 0.20\} \cdot |m|$$
$$p_{max} = \min\{0.5 + 0.3 \cdot h; 0.65\} \cdot |m|$$

Hence, for $h = 0$, modes with $0.2 \leq p/|m| \leq 0.5$ are described, which are typical whispering-gallery modes, while for $h = 1$, typical volume modes with $0.5 \leq p/|m| \leq 0.65$ are pooled. The value $h = 0.5$ encompasses both types. Correspondingly, one can define the character of a mode as

$$h_{mp} := \left(\frac{p}{|m|} - 0.2\right)/0.45$$

As h_{mp} is a purely phenomenological value, the above numbers were chosen by convenience and could be chosen otherwise.

B.2 Axial Continuation of the Transversal Profile

In addition to $\hat{B}_{\|}(x, y)$ being a solution of the transversal field (membrane equation) $\left(\partial_x^2 + \partial_y^2 + k_\perp^2\right) \hat{B}_{\|} = 0$, one can now include the longitudinal field profile $\Gamma(z)$:

$$k_\perp^2 B_z(x, y, z) = k_\perp^2 \Gamma(z) \hat{B}_{\|}(x, y) \equiv -\Gamma(z) \cdot \left(\partial_x^2 + \partial_y^2\right) \hat{B}_{\|}(x, y)$$

Setting

$$k_\perp^2 B_x(x, y, z) = \Gamma'(z) \partial_x \hat{B}_{\|}(x, y)$$
$$k_\perp^2 B_y(x, y, z) = \Gamma'(z) \partial_y \hat{B}_{\|}(x, y)$$

obviously fulfills $\nabla \cdot \vec{B} = 0$, i.e. $\vec{B}(x, y, z)$ is a solution of Maxwell's equations in three dimensions. Thus, in cylindrical coordinates,

$$B_z(r, \varphi, z) = \left(\frac{\Gamma'(z)}{k_\perp^2} \left(\vec{e}_r \partial_r + \vec{e}_\varphi \frac{\partial_\varphi}{r} \right) + \Gamma(z) \vec{e}_z \right) \hat{B}_{\|}(r, \varphi) \qquad \text{(B.16)}$$

B.3 Impedance Boundary Condition of Longitudinal Corrugations

Given a corrugated inner or outer wall at radius R. For simplification, the corrugation cross-section shall be regarded locally as a rectangular shape (width D_φ, depth D_r), which is a good approximation if $D_\varphi \ll 2\pi R$ and $D_r \ll R$. There are M corrugations; thus, the periodicity of corrugations is $2\pi R/M$. The corrugation width shall be smaller than half the cutoff wavelength $D_\varphi < \lambda_\perp/2$. The magnetic field in the rectangular groove can then be described as

$$\hat{B}_{R,z}(r) = \hat{B}_R \cos(k_\perp(r - (R \mp D_r)))$$

where the upper (lower) sign corresponds to the inner (outer) surface, see [Ker96]. According to Eq. B.3a, the corresponding electric field in the corrugations equals $\hat{E}_{R,\varphi}(r) = \frac{-\mathrm{i}\omega}{k_\perp}\hat{B}_R \sin(k_\perp(r-(R\mp D_r)))$. The impedance averaged over all corrugations is then

$$Z_R = \mu_0 \left\langle \frac{\hat{E}_{R,\varphi}(R)}{\hat{B}_{R,z}(R)} \right\rangle = \mu_0 \cdot \frac{D_\varphi \cdot \hat{E}_{R,\varphi}(R) + (\frac{2\pi R}{M} - D_\varphi)\cdot 0}{\frac{2\pi R}{M}\hat{B}_{R,z}(R)} \tag{B.17}$$

$$= \frac{-\mathrm{i}\mu_0\omega}{k_\perp} \cdot \frac{M D_\varphi}{2\pi R} \cdot \tan\left(k_\perp D_r\right) \tag{B.18}$$

Expressing the actual, complicated corrugation structure by an averaged, effective impedance is called the *Surface Impedance Model* (SIM). This is of course only reasonable if the structure is sufficiently smaller than the wavelength. If this is not the case, one will observe resonant/interference effects, which can also be interpreted as coupling to modes with lower azimuthal indices [Ker96], Bragg reflections, and so on. See Section 3.2.

In order to describe the field components within the corrugations later, define a correction term ϖ in accordance with [Ker96], eq. (4.65) and [Avr06], eq. (2.2.15):

$$\varpi := \frac{M_\mathrm{I} k_\perp D_{\varphi,\mathrm{I}}}{2\pi} \cdot \left(k_\perp \cdot \left(\frac{M_\mathrm{I} D_{\varphi,\mathrm{I}}}{2\pi} - D_{r,\mathrm{I}} \right) \cdot \tan^2\left(k_\perp D_{r,\mathrm{I}}\right) - \tan\left(k_\perp D_{r,\mathrm{I}}\right) - k_\perp D_{r,\mathrm{I}} \right) \tag{B.19}$$

For reasonable calculations it is not advisable to neglect the fields inside the corrugations, which would simplify this term to:

$$\varpi' = \frac{M_\mathrm{I} k_\perp D_{\varphi,\mathrm{I}}}{2\pi} \tan\left(k_\perp D_{r,\mathrm{I}}\right) \cdot \left(\frac{M_\mathrm{I} k_\perp D_{\varphi,\mathrm{I}}}{2\pi} \tan\left(k_\perp D_{r,\mathrm{I}}\right) - 2 \right)$$

B.4 Eigenvalue Curves for Cavities with Corrugated Insert

Equation B.17 substituted into Eq. B.14, yields

$$
\begin{aligned}
&-\frac{2\pi}{C_R}\cos\left(k_\perp D_{r,\mathrm{I}}\right)\cdot \hat{B}_{z(,m)}(r) \equiv \hat{B}_m \cdot K_m(k_\perp r) = \\
&= \hat{B}_m \cdot \Big(J_m\left(k_\perp r\right), -N_m\left(k_\perp r\right) \Big)\cdot \\
&\cdot \begin{pmatrix} N'_m\left(k_\perp R_\mathrm{I}\right) & N_m\left(k_\perp R_\mathrm{I}\right) \\ J'_m\left(k_\perp R_\mathrm{I}\right) & J_m\left(k_\perp R_\mathrm{I}\right) \end{pmatrix} \cdot \begin{pmatrix} \frac{2\pi}{C_R}\cos\left(k_\perp D_{r,\mathrm{I}}\right) \\ \frac{M_\mathrm{I} D_{\varphi,\mathrm{I}}}{R_\mathrm{O}}\sin\left(k_\perp D_{r,\mathrm{I}}\right) \end{pmatrix}
\end{aligned}
\tag{B.20}
$$

Consequently, the smooth-wall boundary condition at $r = R_\mathrm{O}$ leads then to the *characteristic equation* for χ_O:

$$
\begin{aligned}
&0 = \left[\partial_\chi K_m(\chi)\right]_{\chi_\mathrm{O}} := \Big(J'_m\left(\chi_\mathrm{O}\right), -N'_m\left(\chi_\mathrm{O}\right) \Big)\cdot \\
&\cdot \begin{pmatrix} N'_m\left(\frac{\chi_\mathrm{O}}{C_R}\right) & N_m\left(\frac{\chi_\mathrm{O}}{C_R}\right) \\ J'_m\left(\frac{\chi_\mathrm{O}}{C_R}\right) & J_m\left(\frac{\chi_\mathrm{O}}{C_R}\right) \end{pmatrix} \cdot \begin{pmatrix} \frac{2\pi}{C_R}\cos\left(\frac{\chi_\mathrm{O} D_{r,\mathrm{I}}}{R_\mathrm{O}}\right) \\ \frac{M_\mathrm{I} D_{\varphi,\mathrm{I}}}{R_\mathrm{O}}\sin\left(\frac{\chi_\mathrm{O} D_{r,\mathrm{I}}}{R_\mathrm{O}}\right) \end{pmatrix}
\end{aligned}
\tag{B.21}
$$

Mind that the derivative of the characteristic function K_m is taken first, then $\chi = \chi_\mathrm{O}$ is set and then χ_I is replaced by χ_O/C_R. The quantity $d_\mathrm{I} := 100 \cdot D_{r,\mathrm{I}}/R_\mathrm{O}$ is used as corrugation parameter (e.g. in CAVITY, see Section C.2.1, as Dd_IM, where the unit and first decimal positions of d_I are used) and can then be given independently of the transversal cavity dimensions. For $D_{r,\mathrm{I}} \approx \lambda/2$, i.e. large impedance, typical values of d_I are between 1 and 3. Equation B.21 can readily be generalized to cavities with impedance corrugations on both walls.

It makes sense to keep the number M_I and transversal geometry of the corrugations constant over z, thus $M_\mathrm{I} D_{\varphi,\mathrm{I}}/2\pi R_\mathrm{I} =: C_\varphi$ can be considered constant. The third factor of K_m in Eq. B.20 is then

$$\frac{2\pi}{C_R}\begin{pmatrix} \cos\left(\frac{\chi_\mathrm{O} D_{r,\mathrm{I}}}{R_\mathrm{O}}\right) \\ C_\varphi \cdot \sin\left(\frac{\chi_\mathrm{O} D_{r,\mathrm{I}}}{R_\mathrm{O}}\right) \end{pmatrix}$$

with C_φ chosen to be between 0 (narrow gaps) and 1 (broad gaps). Concerning roots of K'_m, C_R is then relevant only in the matrix term and K_m is explicitly independent of $D_{\varphi,\mathrm{I}}$. Of course, this is only possible due to the averaging (Eq. B.17); and the condition $D_\varphi < \lambda_\perp/2$ has still to be fulfilled.

For $C_R \to \infty$, the dominant term in the matrix of Eq. B.21 is the Neumann derivative, since the Bessel function and its derivative approach zero and since $N'_m(x) = N_{m-1}(x) - \frac{m}{x} N_m(x)$ grows by a factor of C_R faster than the Neumann function. Thus, in this case, the characteristic function derivative reduces to

$$K'_m(\chi_\mathrm{O}) = J'_m(\chi_\mathrm{O}) \cdot \frac{2\pi}{C_R} \cos\left(\frac{\chi_\mathrm{O} D_{r,\mathrm{I}}}{R_\mathrm{O}}\right) \tag{B.22}$$

Compared to the solutions χ_O in the hollow cavity (or in the case of a smooth insert), the cosine function implies additional eigenvalues ν_p, $p \geq 1$ that are located at

$$\nu_p = \frac{R_\mathrm{O}}{D_{r,\mathrm{I}}} \cdot \left(p - \frac{1}{2}\right)\pi \tag{B.23}$$

These eigenvalues certainly belong to additional TE modes, so-called "inner modes", see Section 3.4, and the other modes will be termed "outer modes" accordingly.

For decreasing C_R, the eigenvalues of inner modes approach zero, while the outer-mode eigenvalues χ_{mp} tend to increase; thus, the respective curves $\chi_{mp}(C_R)$ and $\nu_p(C_R)$ intersect at certain points. The shapes of the curves, and thus the intersections, depend on $\frac{M_{\mathrm{I}} \cdot D_{\varphi,\mathrm{I}}}{R_{\mathrm{O}}}$. The mode numbering becomes now inconsistent in the sense that for different $D_{r,\mathrm{I}}$, M_{I}, or $D_{\varphi,\mathrm{I}}$, the same eigenvalue χ_{O} (up to a very high accuracy) might belong to a different root index p, counting from 0 on upwards. In numerical calculations, this also means that two neighboring roots can be arbitrarily close to each other, so any chosen step size will either increase the computation time (especially for very high-order modes) or increase the probability of missed double roots. A possible eigenvalue (and mode) numbering for a fixed geometry is the following:

- χ^*_{mp} is the p-th root of the characteristic function derivative K'_m.
- χ_{mp} is the root of K_m with an eigenvalue close to the p-th root of the hollow-cavity characteristic function J'_m.

Each "perturbed" eigenvalue curve $\chi^*_{mp}(C_R)$ has then intervals where it is an outer mode and intervals where it is an inner mode. In most of the first case, $\frac{\mathrm{d}\chi}{\mathrm{d}C_R} < 0$, while in the second case $\frac{\mathrm{d}\nu}{\mathrm{d}C_R} > 0$. If the radius of the insert varies along z, one obtains the differentials

$$\frac{\mathrm{d}\chi}{\mathrm{d}z} = \frac{\mathrm{d}\chi}{\mathrm{d}C_R} \cdot \frac{\mathrm{d}C_R}{\mathrm{d}z}$$
$$\frac{\mathrm{d}\nu}{\mathrm{d}z} = \frac{\mathrm{d}\nu}{\mathrm{d}C_R} \cdot \frac{\mathrm{d}C_R}{\mathrm{d}z}$$

In order to enhance the diffractive quality factor Q_{diff} of a mode, it has to come closer to cutoff, i.e. its eigenvalue has to be increased with increasing axial coordinate z [IKP96, Ker96]. This leads to the favourable conditions

$$\frac{\mathrm{d}C_R}{\mathrm{d}z} < 0 \quad \text{(for outer modes)}$$
$$\frac{\mathrm{d}C_R}{\mathrm{d}z} > 0 \quad \text{(for inner modes)}$$

The opposite sign stands for decreasing values of Q_{diff}. In a coaxial gyrotron with narrowing coaxial insert, to select one mode to be enhanced (or unaffected) and its competitors to be disrupted, one could choose a main mode that is in an inner-mode interval, while the main competitors are outer modes. However, the electromagnetic field of modes with sufficiently large $\frac{\mathrm{d}\chi}{\mathrm{d}z}$ is concentrated on the inner rod and hence these modes have large ohmic loading on the insert, i.e. Q_Ω is small, see Section 3.4 [IKP96, Avr06]. Thus, the optimum choice for an operating mode is where $\frac{\mathrm{d}\chi}{\mathrm{d}z} \approx 0$, while for its main competitors either $\frac{\mathrm{d}\chi}{\mathrm{d}z} > 0$ or $\frac{\mathrm{d}\chi}{\mathrm{d}z} < 0$ holds.

B.5 Characteristic Matrix of a Misaligned Cavity with Two Impedance Boundaries

The most general characteristic matrix in the frame of this work is the one of a misaligned cavity with impedance corrugations on the inner wall and on the insert. Its elements follow directly from Eq. B.12, using the same derivation as for Eq. 3.7, resulting in:

$$\begin{aligned} K_{qm}(k_\perp) =& J_{q-m}\left(k_\perp D_{\mathrm{IO}}\right) \cdot \\ & \cdot \begin{pmatrix} \cos\left(k_\perp D_{r,\mathrm{O}}\right), & \frac{M_\mathrm{O} D_{\varphi,\mathrm{O}}}{2\pi R_\mathrm{O}} \sin\left(k_\perp D_{r,\mathrm{O}}\right) \end{pmatrix} \cdot \begin{pmatrix} J'_q\left(k_\perp R_\mathrm{O}\right) & -N'_q\left(k_\perp R_\mathrm{O}\right) \\ J_q\left(k_\perp R_\mathrm{O}\right) & -N_q\left(k_\perp R_\mathrm{O}\right) \end{pmatrix} \cdot \\ & \cdot \begin{pmatrix} N'_m\left(k_\perp R_\mathrm{I}\right) & N_m\left(k_\perp R_\mathrm{I}\right) \\ J'_m\left(k_\perp R_\mathrm{I}\right) & J_m\left(k_\perp R_\mathrm{I}\right) \end{pmatrix} \cdot \begin{pmatrix} \cos\left(k_\perp D_{r,\mathrm{I}}\right) \\ \frac{M_\mathrm{I} D_{\varphi,\mathrm{I}}}{2\pi R_\mathrm{I}} \sin\left(k_\perp D_{r,\mathrm{I}}\right) \end{pmatrix} \end{aligned} \quad \text{(B.26)}$$

In this formalism, the misalignment (first term), the outer wall (terms 2 and 3) and the insert wall (terms 4 and 5), as well as the impedance corrugations (terms 2 and 5) are visibly separated from each other. Equations such as to determine the field amplitudes (Eq. 3.9) or the normalization (Eq. 3.12) remain valid.

Mind that terms 2 and 5 are not adjacent (and cannot be made so) and thus it is *not* possible to merge them into a common matrix that would describe an effective corrugation assigned to one of the walls exclusively.

B.6 Normalization Constant and Energy Content of TE Modes with Impedance Boundary Conditions

A measure for the relative interaction strength of a TE mode is the integrated electric field strength (here, $\vec{E} \equiv E_\perp$) over the cavity cross-section:

$$U_\perp^2 = \int\limits_{S_\perp} \left|\vec{E}\right|^2 \mathrm{d}x\mathrm{d}y = \int\limits_0^{2\pi} \mathrm{d}\varphi \int\limits_{R_\mathrm{I}}^{R_\mathrm{O}} \mathrm{d}r\, r \left(|E_\varphi|^2 + |E_r|^2\right) \tag{B.27}$$

Using the definitions of E_φ and E_r (Eqs. B.7a, B.7b), and using $\chi = k_\perp r$, $\chi_\mathrm{I} = k_\perp R_\mathrm{I}$, $\chi_\mathrm{O} = k_\perp R_\mathrm{O}$ (the cutoff wavevector $k_\perp$ is known from the above eigenvalue considerations), one obtains

$$U_\perp^2 = \frac{2\pi c^2}{k_\perp^2} \int\limits_{\chi_\mathrm{I}}^{\chi_\mathrm{O}} \mathrm{d}\chi \left(\chi \left(\frac{\mathrm{d}\hat{B}_z}{\mathrm{d}\chi}\right)^2 + \frac{m^2}{\chi}\hat{B}_z^2\right)$$

Since $\hat{B}_z$ is a linear combination of J_m and N_m and thus fulfills the Bessel differential equation, the integrand can be rewritten as

$$\begin{aligned}
&\chi\left(\hat{B}_z'\right)^2 + \frac{m^2}{\chi}\hat{B}_z^2 \\
=&\chi\left(\hat{B}_z'\right)^2 + \frac{\hat{B}_z}{\chi}\left(\chi^2\hat{B}_z'' + \chi\hat{B}_z' + \chi^2\hat{B}_z\right) - \\
&- \hat{B}_z'\left(m^2\hat{B}_z - \chi^2\hat{B}_z - \chi\hat{B}_z' - \chi^2\hat{B}_z''\right) \\
=&\left(\chi\hat{B}_z'^2 + \chi\hat{B}_z\hat{B}_z'' + \hat{B}_z\hat{B}_z'\right) + \left(\chi\hat{B}_z^2 - m^2\hat{B}_z\hat{B}_z' + \chi^2\hat{B}_z\hat{B}_z'\right) + \\
&+ \left(\chi\hat{B}_z'^2 + \chi^2\hat{B}_z'\hat{B}_z''\right) \\
=&\left(\chi\hat{B}_z\hat{B}_z' - \frac{1}{2}(m^2 - \chi^2)\hat{B}_z^2 + \frac{1}{2}\chi^2\hat{B}_z'^2\right)'
\end{aligned}$$

Therefore,

$$U_\perp^2 = \frac{2\pi c^2}{k_\perp^2} \cdot \left[\chi \hat{B}_z \hat{B}_z' - \frac{1}{2}(m^2 - \chi^2)\hat{B}_z^2 + \frac{1}{2}\chi^2 \hat{B}_z'^2 \right]_{\chi_\mathrm{I}}^{\chi_\mathrm{O}} \tag{B.28}$$

The derivative $\hat{B}_z'$ can be substituted using the impedance boundary condition Eq. B.11:

$$\begin{aligned} U_\perp^2 =& \frac{2\pi c^2}{k_\perp^2} \left[\frac{\chi Z k_\perp}{\mathrm{i}\mu_0\omega} \hat{B}_z^2 - \frac{1}{2}(m^2 - \chi^2)\hat{B}_z^2 + \frac{1}{2}\left(\frac{\chi Z k_\perp}{\mathrm{i}\mu_0\omega} \hat{B}_z \right)^2 \right]_{\chi_\mathrm{I}}^{\chi_\mathrm{O}} \\ =& \frac{\pi c^2}{k_\perp^2} \hat{B}_z^2(\chi_\mathrm{O}) \left(\chi_\mathrm{O}^2 - m^2 - 1 + \left(\frac{\chi_\mathrm{O} Z_\mathrm{O} k_\perp}{\mathrm{i}\mu_0\omega} + 1 \right)^2 \right) - \\ & - \frac{\pi c^2}{k_\perp^2} \hat{B}_z^2(\chi_\mathrm{I}) \left(\chi_\mathrm{I}^2 - m^2 - 1 + \left(\frac{\chi_\mathrm{I} Z_\mathrm{I} k_\perp}{\mathrm{i}\mu_0\omega} + 1 \right)^2 \right) \end{aligned}$$

With longitudinal corrugations (Eq. B.17) on the insert and with smooth outer wall, the constant reads

$$U_\perp^2 = \frac{\pi c^2}{k_\perp^2} \left(\hat{B}_z^2(R_\mathrm{O}) \cdot (\chi_\mathrm{O}^2 - m^2) - \hat{B}_z^2(R_\mathrm{I}) \cdot (\chi_\mathrm{I}^2 - m^2 + \varpi) \right) \tag{B.29}$$

where $\hat{B}_z(r)$ can be calculated from Eq. B.20. As one can easily derive from Eq. B.27 and from Section B.2, the actual stored field energy in a cylindrical cavity of length $L_\mathrm{O} \equiv z_2 - z_1$ and longitudinal field profile $\Gamma(z)$ is

$$W = \frac{\varepsilon_0}{2} U_\perp^2 \int_{z_1}^{z_2} \Gamma(z)^2 \mathrm{d}z \tag{B.30}$$

B.7 Ohmic Wall Loading and Ohmic Quality Factor

See [Jac99], 5.18 and 8.1. The magnetizing field $\vec{H}$ at either wall of the coaxial cavity, close to the surface, is considered. One can choose the local coordinate system such that the origo lies in the surface of the wall, $\vec{e}_x$

points outwards (from the wall material into the vacuum), $\vec{e}_y$ is tangential to the wall, and $\vec{e}_z$ points along the cavity axis. Neglecting the sources of the electric displacement field and Maxwell's displacement current and introducing the electric conductivity σ (with $\vec{j} = \sigma \vec{E}$), Maxwell's (macroscopic) equations read

$$\begin{aligned}
\nabla \cdot \vec{D} &= 0 \\
\nabla \cdot \vec{B} &= 0 \\
\nabla \times \vec{H} &= \vec{j} = \sigma \vec{E} \\
\nabla \times \vec{E} &= -\partial_t \vec{B}
\end{aligned}$$

Using simple vector algebra, one obtains for the magnetizing field $\vec{H}$ (and analogously for $\vec{A}$, $\vec{E}$, $\vec{j}$):

$$\Delta \vec{H} = \sigma\mu \, \partial_t \vec{H}$$

where $\mu = \mu_0 \mu_{\text{rel}}$. The field in x-direction vanishes; however, the change of the other two field components in x-direction is strongest. Hence, one obtains for the tangential magnetizing field $H_{\text{tang}} = |H_y \vec{e}_y + H_z \vec{e}_z|$ (not to be confused with perpendicular component $H_\perp = H_x + \mathrm{i} H_y$):

$$\partial_x^2 H_{\text{tang}} \approx \sigma\mu \, \partial_t H_{\text{tang}}$$

The harmonic ansatz $H_{\text{tang}}(x, y, z, t) = \hat{H}_{\text{tang}}(y, z) e^{\mathrm{i}\omega t} e^{-\mathrm{i}kx}$ returns

$$H_{\text{tang}} = \hat{H}_{\text{tang}}(y, z) e^{\mathrm{i}\omega t} e^{\mathrm{i}x/D_{\text{S}}} e^{x/D_{\text{S}}}$$

for the field inside the wall, where $D_{\text{S}} := \sqrt{\frac{2}{\sigma\mu\omega}}$ is the skin depth.

The corresponding current (which is perpendicular to the magnetic field direction) results from Ampère's law, and the tangential electric field (first-order correction) results from the definition of σ:

$$\begin{aligned} j_{\text{tang}} &= \frac{\sqrt{2}}{D_{\text{S}}} \hat{H}_{\text{tang}} \cdot e^{x/D_{\text{S}}} e^{\mathrm{i}(\omega t + x/D_{\text{S}} + 5\pi/4)} \\ E_{\text{tang}} &= \frac{\sqrt{2}}{\sigma D_{\text{S}}} \hat{H}_{\text{tang}} \cdot e^{x/D_{\text{S}}} e^{\mathrm{i}(\omega t + x/D_{\text{S}} + 5\pi/4)} \end{aligned}$$

The work per unit volume on the electrons is then the time-average over the product of the real parts of both (with $\hat{H}^2_{\text{tang}} = \hat{H}^2_y + \hat{H}^2_z$):

$$\frac{\mathrm{d}P_\Omega}{\mathrm{d}V} = \langle j_{\text{tang}} E_{\text{tang}} \rangle = \frac{\mu\omega}{2} \left(\hat{H}^2_y + \hat{H}^2_z \right) e^{2x/D_{\text{S}}}$$

Integration from $x = -\infty$ to 0 yields the total ohmic losses per surface element, i.e. the ohmic loading (see e.g. [KBT04], eq. 5.3.), namely

$$w_\Omega \equiv \frac{\mathrm{d}P_\Omega}{\mathrm{d}S} = \frac{\mu\omega}{4} D_{\text{S}} \cdot \left(\hat{H}^2_y + \hat{H}^2_z \right)$$

Using Eq. B.16 and $\omega = 2\pi\mu_{\text{rel}} c/\lambda$, one obtains for the ohmic loading on an axisymmetric boundary with radius R from a TE mode:

$$w_{\Omega,m}(R) = \frac{\pi c}{2\mu_0} \frac{D_{\text{S}}}{\lambda} \left(\left(\frac{\Gamma'(z)}{k_\perp} \cdot \frac{m}{k_\perp R} \right)^2 + \Gamma(z)^2 \right) \cdot \hat{B}^2_{\parallel,m}(R) \qquad \text{(B.31)}$$

The field component $\hat{B}_{\parallel,m}(R)$ can be estimated from Eq. B.14. For a typical cavity with length L_{O} one can estimate the ratio between the squared terms as

$$\frac{\mathrm{d}\Gamma/\mathrm{d}(k_\perp z)}{\Gamma} \cdot \frac{m}{\chi} \approx \frac{\lambda}{\pi L_{\text{O}}} \cdot \frac{m}{\chi} \ll 1$$

which leads to the conclusion that the loading by the z-component of the field is much larger than that by the φ-component.

The ohmic quality factor $Q_\Omega := \frac{\omega W}{P_\Omega}$ of a TE mode in a cylindrical cavity can now be obtained using Eq.s B.29 and B.31 (with $\Gamma(z) \equiv 1$):

$$\begin{aligned} Q_\Omega &= \omega \cdot \frac{\frac{\varepsilon_0}{2} U_\perp^2}{2\pi \left(R_\mathrm{O} w_\Omega(R_\mathrm{O}) + R_\mathrm{I} w_\Omega(R_\mathrm{I}) \right)} \\ &= \frac{R_\mathrm{O}}{\chi_\mathrm{O}} \cdot \frac{\hat{B}_z^2(R_\mathrm{O}) \left(\chi_\mathrm{O}^2 - m^2 \right) - \hat{B}_z^2(R_\mathrm{I}) \left(\chi_\mathrm{I}^2 - m^2 + \varpi \right)}{D_\mathrm{S,O}\, \chi_\mathrm{O}\, \hat{B}_z^2(R_\mathrm{O}) + D_\mathrm{S,I}\, \chi_\mathrm{I}\, \hat{B}_z^2(R_\mathrm{I})} \end{aligned} \tag{B.32}$$

One can readily define the individual contributions of insert wall Q_I and cavity wall Q_O.

If no insert is present, $\hat{B}_z(R_\mathrm{I})$ vanishes and the remaining field amplitudes cancel out; and one obtains the well-known result ([KBT04] (5.11))

$$Q_{\Omega,\mathrm{hollow}} = \frac{R_\mathrm{O}}{D_\mathrm{S,O}} \cdot \left(1 - \frac{m^2}{\chi_\mathrm{O}^2} \right) \tag{B.33}$$

Applying the general relation for powers and quality factors to the stored energy and frequency, $P_\Omega Q_\Omega = \omega W$, leads from Eq. B.33 with the definition of skin depth to

$$P_\Omega = \sqrt{\frac{2\,\varepsilon_0\,\omega^3}{\sigma}} \cdot \frac{W/\chi_\mathrm{O}}{1 - (m/\chi_\mathrm{O})^2} \tag{B.34}$$

Since the skin depth is typically less than 1 µm, hence smaller than possible corrugations, the metallic surfaces can be regarded as perfect conductors when the mode eigenvalues are calculated. Thus, nonideal surfaces do not impede the applicability of the SIM (Section B.3), but one might need to calculate the ohmic loading inside the corrugations in more detail, e.g. for each kind of the surfaces (bottom, top and side) separately [DZ04], denoted ϖ in this work. In realistic scenarios, however, the corrugations will not be perfectly formed, which slightly influences both ohmic loading and mode eigenvalues.

B.8 Interaction of Modes and Particles

The following considerations are based on [Bor91]. The hollow electron beam, aligned to the cavity wall, is thought to consist of n_b beamlets with guiding centers $x_{\perp i_\mathrm{b}} \equiv r_\mathrm{b} e^{\mathrm{i}\varphi_{i_\mathrm{b}}}$. Each beamlet consists of n_C (macro-)particles, located at $x_{\perp i_\mathrm{C}} \equiv r_\mathrm{C} e^{\mathrm{i}\varphi_{i_\mathrm{C}}}$ with respect to the guiding center. In the following, only one particle will be considered and the counting indices i will be omitted. Then, the location of the particle (e.g. in the outer-wall coordinate system) is

$$x_\mathrm{b} + x_\mathrm{C} = x_\perp = r e^{\mathrm{i}\varphi} \tag{B.35}$$

and its momentum is tangential to the circular (or helical) path:

$$p_\perp = \hat{p} e^{\mathrm{i}(\varphi_\mathrm{b} + \varphi_\mathrm{C} + \frac{\pi}{2})} \tag{B.36}$$

One can now return to Eqs. A.15b-A.17, i.e. to the Lorentz force on individual particles (or macroparticles):

$$\frac{\mathrm{d}p_\perp}{\mathrm{d}t} = \mathrm{q_e} E_\perp + \mathrm{i}\mathrm{q_e} v_\parallel B_\perp - \mathrm{i}\mathrm{q_e} v_\perp B_\parallel \tag{B.37}$$

$$\frac{\mathrm{d}p_\parallel}{\mathrm{d}t} = \mathrm{q_e} E_\parallel + \mathrm{q_e}\mathrm{Im}(v_\perp^* B_\perp) \tag{B.38}$$

$$\frac{\mathrm{d}W}{\mathrm{d}t} = \mathrm{q_e} v_\parallel E_\parallel + \mathrm{q_e}\mathrm{Re}(v_\perp^* E_\perp) \tag{B.39}$$

For TE modes one may neglect the transverse magnetic fields due to the RF waves, leaving only the static external, axial field to be accounted for (termed “adiabatic approximation” of the magnetic field). Furthermore, the acceleration due to $E_\parallel$ can be separated from the equations. Using Eq. A.8 in complex notation and $W =: \mathrm{Re}(W_\perp)$ yields

$$
\begin{aligned}
\frac{\mathrm{d}p_\perp}{\mathrm{d}t} =& \mathrm{q_e} E_\perp - \mathrm{i}\mathrm{q_e} v_\parallel \frac{x_\perp}{2} \partial_\parallel B_\parallel - \mathrm{i}\mathrm{q_e} v_\perp B_\parallel \\
\frac{\mathrm{d}p_\parallel}{\mathrm{d}t} =& - \mathrm{q_e} \mathrm{Im}(v_\perp^* \frac{x_\perp}{2} \partial_\parallel B_\parallel) \\
\frac{\mathrm{d}W_\perp}{\mathrm{d}t} =& \mathrm{q_e} v_\perp^* E_\perp
\end{aligned}
$$

With $v = \frac{p}{\gamma \mathrm{m_e}}$ and $\Omega_\mathrm{C} = -\frac{\mathrm{q_e} B_\parallel}{\gamma \mathrm{m_e}}$ one obtains:

$$
\begin{aligned}
\mathrm{i}\frac{\mathrm{d}p_\perp}{\mathrm{d}t} =& \mathrm{i}\mathrm{q_e} E_\perp + \frac{\mathrm{q_e} \partial_\parallel B_\parallel}{2\gamma \mathrm{m_e}} p_\parallel x_\perp - \Omega_\mathrm{C} p_\perp \\
\frac{\mathrm{d}p_\parallel}{\mathrm{d}t} =& - \frac{\mathrm{q_e} \partial_\parallel B_\parallel}{2\gamma \mathrm{m_e}} \mathrm{Im}(p_\perp^* x_\perp) \\
\frac{\mathrm{d}W_\perp}{\mathrm{d}t} =& \frac{\mathrm{q_e}}{\gamma \mathrm{m_e}} p_\perp^* E_\perp
\end{aligned}
$$

The time-derivative can now be executed on the perpendicular momentum (Eq. B.36). With $\frac{\mathrm{d}X}{\mathrm{d}t} \equiv \dot{X}$ and inserting the expressions Eq. B.35 and B.36, one finds:

$$
\left(\mathrm{i}\frac{\mathrm{d}}{\mathrm{d}t} - \dot{\varphi_\mathrm{b}} - \dot{\varphi_\mathrm{C}} + \Omega_\mathrm{C} \right) \hat{p} = \mathrm{q_e} E_\perp e^{-\mathrm{i}(\varphi_\mathrm{b} + \varphi_\mathrm{C})} + \frac{\mathrm{q_e} \partial_\parallel B_\parallel}{2\gamma \mathrm{m_e}} p_\parallel r e^{\mathrm{i}\varphi} \tag{B.40}
$$

$$
\frac{\mathrm{d}p_\parallel}{\mathrm{d}t} = - \frac{\mathrm{q_e} \partial_\parallel B_\parallel}{2\gamma \mathrm{m_e}} \cos(\varphi - \varphi_\mathrm{b} - \varphi_\mathrm{C}) \tag{B.41}
$$

$$
\mathrm{i}\frac{\mathrm{d}W_\perp}{\mathrm{d}t} = \frac{\mathrm{q_e} \hat{p}}{\gamma \mathrm{m_e}} E_\perp e^{-\mathrm{i}(\varphi_\mathrm{b} + \varphi_\mathrm{C})} \tag{B.42}
$$

The transversal field is now expressed as a sum (linear superposition) of eigenmodes, using an unitless axial envelope $\Gamma_{mp}(z)$ from the longitudinal field profile considerations, Eq. B.8 for the transversal field profile, and Eq. B.28 for normalization. Factorization of the longitudinal and transversal terms is an appropriate approximation only if the tapering $\partial_z R(z)$ is small.

$$E_{\perp}(r, \varphi, z, t) = \\ - \sum_{mp} k_{mp} \frac{\hat{B}_{\parallel,m-1}(k_{mp} r)/\hat{B}_{J,mp}}{U_{\perp,mp}/(\omega_{mp}\hat{B}_{J,mp}/k_{mp}^2)} U_{\perp,mp} \Gamma_{mp}(z) e^{-\mathrm{i}m\varphi} e^{\mathrm{i}\omega_{mp} t} \tag{B.43}$$

Written in this form, it is obvious that the double fraction in the sum (second factor) depends only on the given geometry; numerator and denominator are well-defined also for large k_{mp}. The denominator is commonly denoted $1/C_{mp}$ (as in [Bor91, Ker96]). The third and fourth factors are equivalent to V_{mp} and $\nabla \times \psi_{mp}$ in [Bor91], respectively, since $\psi_{mp} \sim B_{\parallel}$.

For a hollow cavity with smooth walls, the numerator (see Eq. B.6) and denominator (see Eq. B.29) are:

$$\begin{aligned} \Upsilon_{mp}^{(\mathrm{num})} &= J_{m-1}\left(\frac{\chi_{mp} r}{R_{\mathrm{O}}}\right) \\ \Upsilon_{mp}^{(\mathrm{denom})} &= \sqrt{\pi\left(\chi_{mp}^2 - m^2\right)}\, J_m(\chi_{mp}) \end{aligned} \tag{B.44}$$

Inserting the mode decomposition Eq. B.43 into Eq. B.40/B.42 and preparing the terms for the next steps yields:

$$\begin{aligned} &\left(\mathrm{i}\frac{\mathrm{d}}{\mathrm{d}t} - \dot{\varphi_{\mathrm{b}}} - \dot{\varphi_{\mathrm{C}}} + \Omega_{\mathrm{C}}\right)\hat{p} - \frac{\mathrm{q_e}\partial_{\parallel} B_{\parallel}}{2\gamma \mathrm{m_e}} p_{\parallel} r e^{\mathrm{i}\varphi} = \\ &= -\,\mathrm{q_e} \sum_{mp} \frac{k_{mp} U_{\perp,mp} \Gamma_{mp}(z) e^{\mathrm{i}\omega_{mp} t}}{U_{\perp,mp}/(\omega_{mp}/k_{mp}^2)} e^{-\mathrm{i}(\varphi + m\varphi_{\mathrm{b}} + \varphi_{\mathrm{C}})} \cdot \\ &\quad \cdot \hat{B}_{\parallel,m-1}(k_{mp} r) e^{\mathrm{i}(1-m)(\varphi - \varphi_{\mathrm{b}})} \end{aligned} \tag{B.45}$$

$$\begin{aligned} \mathrm{i}\frac{\mathrm{d}W_{\perp}}{\mathrm{d}t} &= \frac{-\mathrm{q_e}\hat{p}}{\gamma \mathrm{m_e}} \sum_{mp} \frac{k_{mp} U_{\perp,mp} \Gamma_{mp}(z) e^{\mathrm{i}\omega_{mp} t}}{U_{\perp,mp}/(\omega_{mp}/k_{mp}^2)} e^{-\mathrm{i}(\varphi + m\varphi_{\mathrm{b}} + \varphi_{\mathrm{C}})} \cdot \\ &\quad \cdot \hat{B}_{\parallel,m-1}(k_{mp} r) e^{\mathrm{i}(1-m)(\varphi - \varphi_{\mathrm{b}})} \end{aligned} \tag{B.46}$$

Recalling that positive and negative integers are interchangeable as indices of Bessel and Neumann functions, the term after the multiplication sign can be expanded in a sum according to Graf's theorem [AS64]:

$$\hat{B}_{\|,m-1}(k_{mp}r)e^{\mathrm{i}(1-m)(\varphi-\varphi_\mathrm{b})} = \sum_{(s-1)=-\infty}^{+\infty} \hat{B}_{\|,m-s}(k_{mp}r_\mathrm{b})J_{s-1}(k_{mp}r_\mathrm{C})e^{\mathrm{i}(s-1)(\pi-\varphi_\mathrm{C})} \tag{B.47}$$

Inserting and sorting the exponential terms yields:

$$\left(\mathrm{i}\frac{\mathrm{d}}{\mathrm{d}t} - \dot{\varphi_\mathrm{b}} - \dot{\varphi_\mathrm{C}} + \Omega_\mathrm{C}\right)\hat{p} - \frac{\mathrm{q_e}\partial_\| B_\|}{2\gamma \mathrm{m_e}} p_\| r e^{\mathrm{i}\varphi} = \\ = \mathrm{q_e}\sum_{mp} \frac{k_{mp}U_{\perp,mp}\Gamma_{mp}(z)e^{-\mathrm{i}m\varphi_\mathrm{b}}}{U_{\perp,mp}/(\omega_{mp}/k_{mp}^2)} \\ \sum_{(s-1)} \hat{B}_{\|,m-s}(k_{mp}r_\mathrm{b})J_{s-1}(k_{mp}r_\mathrm{C})e^{\mathrm{i}\pi s}e^{\mathrm{i}(\omega_{mp}t-s\varphi_\mathrm{C})}e^{-\mathrm{i}\varphi} \tag{B.48}$$

$$\mathrm{i}\frac{\mathrm{d}W_\perp}{\mathrm{d}t} = \frac{\mathrm{q_e}\hat{p}}{\gamma \mathrm{m_e}}\sum_{mp} \frac{k_{mp}U_{\perp,mp}\Gamma_{mp}(z)e^{-\mathrm{i}m\varphi_\mathrm{b}}}{U_{\perp,mp}/(\omega_{mp}/k_{mp}^2)} \\ \sum_{(s-1)} \hat{B}_{\|,m-s}(k_{mp}r_\mathrm{b})J_{s-1}(k_{mp}r_\mathrm{C})e^{\mathrm{i}\pi s}e^{\mathrm{i}(\omega_{mp}t-s\varphi_\mathrm{C})}e^{-\mathrm{i}\varphi} \tag{B.49}$$

The dimensionless coupling factor for each mode (for one chosen s) consists of the absolute parts of the right side of Eq. B.48, normalized to the maximum field strength at the respective axial position:

$$G_{mps}(r_\mathrm{b}) = \frac{\chi_{mp}\hat{B}_{\|,m-s}(k_{mp}r_\mathrm{b})}{U_{\perp,mp}/(\omega_{mp}/k_{mp}^2)} \cdot J_{s-1}(k_{mp}r_\mathrm{C}) \tag{B.50}$$

If only one harmonic and modes in a small k_{mp} range are considered, the second factor may be omitted. The beam radius for optimum coupling is at the first extremum of $\partial_{r_\mathrm{b}}({G_{mps}}^2)$, i.e. where

$$r_{\mathrm{b,opt}} = \frac{\chi_{m-s,1}}{\chi_{mp}} \cdot R_\mathrm{O} \tag{B.51}$$

One can also set $r_\mathrm{b} = R_\mathrm{I}$ or $r_\mathrm{b} = R_\mathrm{O}$ to compare the coupling of the beam with the "wall couplings", i.e. with the respective ohmic loss.

B.9 Tunability Step Size

As discussed for example in [FAG+15], the eigenvalues of typical high-order cavity modes (there: $31 \le m \le 43$, $11 \le p \le 17$ with a relative error of less than $0.3\ \%$) can be linearly approximated by

$$\chi_{mp} \approx 3.415p + 1.268m - 5.11$$

During frequency-step tuning, the excited modes are in most cases azimuthal neighbors, hence

$$\Delta\chi \approx 1.268$$

With the relations $\chi = 2\pi k_\perp R_\mathrm{O}$ and $\omega = ck$, one obtains for steps between modes near cutoff:

$$R_\mathrm{O} \cdot \Delta\omega \approx 61\ \mathrm{mm} \cdot \mathrm{GHz} \tag{B.52}$$

Reference [ZT05] gives several examples for step-tuned gyrotrons in its section 3, e.g. $58\ \mathrm{mm} \cdot \mathrm{GHz}$ for a $\mathrm{TE}_{22,6}$ mode gyrotron, $59\ \mathrm{mm} \cdot \mathrm{GHz}$ for a $\mathrm{TE}_{25,10}$ mode gyrotron, $57\ \mathrm{mm} \cdot \mathrm{GHz}$ for a $\mathrm{TE}_{31,8}$ mode gyrotron, and $62\ \mathrm{mm} \cdot \mathrm{GHz}$ for a coaxial $\mathrm{TE}_{34,19}$ mode gyrotron.

C Description of the Used Software

C.1 Numerical Approaches

In order to solve an electrodynamics problem numerically, one can base the calculations on formulas at any level of abstractness and on any appropriate approximations. A typical approach for the calculation of gyrotron interaction are Particle-In-Cell-Codes using Lorentz and Maxwell's equations directly: the electrons are treated as point-like multiparticles (having the charge of a large, but fixed number of electrons), moving through a mesh that has electric and magnetic fields assigned to each node. The Lorentz force acting on a multiparticle at a defined position is interpolated from the fields of the surrounding nodes, while the field at each node is, if not fixed or externally imposed, calculated via the Maxwell's equations with the multiparticles as sources. Alternatively, one can use approximations and simplifications that are typical for gyrotrons, such as

- axisymmetry, thus separation of the azimuthal coordinate;
- small tapering, thus separation of the axial coordinate and neglect of mode conversion;
- description of the electron trajectories as small orbits around defined guiding centers;
- slowly varying field amplitudes $|\partial X/\partial t| \ll \omega \cdot |X|$, thus decoupling them from the fast interaction regime;
- conservation of the electrons' axial momentum; and/or
- disregard of RF magnetic fields.

C.2 Codes Used / Developed

C.2.1 CAVITY Package

CAVITY is a KIT in-house package of twelve routines for design, simulation and optimization of gyrotron cavities, both hollow and coaxial. Its core functions have been written between 1986 (SELFC) and 1997 (SELFT), with improvements until 2011. The routines (or applications) are independent executables, however linked by interface files which are used as output by some, as input by other routines. The package is almost exclusively written in FORTRAN-77, with one exception being the Tcl/Tk user interface (by Harig, Kern, Illy 1997), which can also serve as a developer interface. The CAVITY routines are not parallelized. See also [Ker96], section 5.2, and Fig. 2.5. The following routines have been used for this work.

MAXPO (Iatrou, Kern 1996)

MAXPO estimates the maximum possible output power of a gyrotron for a set of general design parameters such as frequency f, efficiency η_{tot}, beam radius r_{b}, dimensionless variable μ, maximum emitter current density $j_{\mathrm{E,max}}$, and so on.

GEOMT (Kern 1993)

GEOMT is the main tool to define a geometry. Cavity wall and, if desired, coaxial insert are defined by an axial profile of straight lines with inclination angles, perhaps joined by parabolic smoothings. Parts of the insert can be corrugated, defined either by the corrugation depth `DxyM` or by the corrugation parameter `WxyM`, with `x` and `y` being the ones place and tenths place of d_{I} and $w := C_{\varphi} \tan(k_{\perp} D_r)$, respectively, and by `s/t`$= C_{\varphi}$ separately. Alternatively, the auxiliary package GEOFREE can be used for geometry input from external files, e.g. for geometries with arbitrary contour.

SCNCHI (Borie 1993)

For a given geometry, a given z range and a given r_b range, SCNCHI finds all modes within a certain frequency range (or, eigenvalue range, therefore the name *scan chi*) and above a certain relative coupling with respect to the desired main mode. Optionally, the coupling factor $G(z, r_\mathrm{b})$ for a given mode can be calculated or a scan over the C_R-χ plane can be performed to find relevant modes.

In coaxial cavities, SCNCHI can not find all roots due to numerical overflow of the Neumann functions. Therefore, unpredictable shifts in the radial index appear that are dependent on the axial coordinate. Thus, the calculated coupling is in most cases incorrect. Due to the memory-saving programming style of the CAVITY routines, this could not be fixed without major rewriting of the code.

CHIMP (Borie 1993)

Within a given eigenvalue and radii ratio range, this routine finds the eigenvalue curves χ_{mp} using detailed input of insert impedance corrugation and, if necessary, rippled-wall corrugation. Corrugation variables can be swept over. For a given mode possible conversions can be found. The correctness of these results, however, relies on the proper identification of mode eigenvalues and mode indices, which is not always given for coaxial cavities due to the same reasons as in SCNCHI.

COLDC (Borie, Dumbrajs, Kern 1989-1993)

COLDC calculates cold-cavity, i.e. excitation-free, axial field profiles $\Gamma(z)$, eigenfrequencies, and quality spectra for the modes found by SCNCHI. The field profile is found in Vlasov approximation [VZO$^+$69], i.e. considering a waveguide with only small irregularities.
Several options are available to find reasonable field profiles, which is especially relevant if a completely new geometry is investigated. As with

the following routines, consistent assignment of eigenvalues and mode indices is required for accurate calculations.

GYMOT (Borie, Kern 1989-1993)

Using the cold-cavity (or, at a later stage, self-consistent stationary or self-consistent) field profiles, the equation of motion of traversing electrons is solved such that required interaction parameters (B_{O}, I_{b}, W_{e}, g_{O}) and possible power and efficiency can be determined. Possible output are starting current curves over U_{b} or B_{O}. GYMOT has proven to be a very powerful tool in the cavity design phase of this work.

SELFC (Borie, Kern 1986-1993)

SELFC calculates stationary self-consistent single-mode solutions of equation of motion and field profile (in Vlasov approximation). As it includes both the influence of the field on the electrons and of the electrons on the field, it gives more accurate results than COLDC and GYMOT alone. SELFC also calculates ohmic loading on insert and cavity wall.

MCONV (Braz (Höchtl), Wien, Kern 1993-1994)

This is a tool for calculation of mode conversion within a z-range of a given cavity. It is mainly used to confirm that the main mode does not convert to other modes in the uptaper section.

SELFT (Kern 1995, 2008-2011)

SELFT provides time-dependent self-consistent multimode simulations of the gyrotron interaction. As such it is the core routine of the CAVITY package. Apart from numerical parameters (length of time steps, number of macroelectrons) the user can define a startup scenario and choose the modes to be included. The code uses output data from other routines such as GEOMT for the cavity geometry and SELFC/SELFT for the initial axial field profiles. Typical output are time-dependent power levels, spectra and axial profiles of the individual modes. Not implemented are thick or misaligned beams and ohmic loading considerations. During interaction simulations with SELFT, an unexpectedly strong correlation between corrugation depth and output power as well as between velocity spread and output power drop was found.

C.2.2 EURIDICE Package (Avramidis 2006-2016)

The EURIDICE package [APIV12, Avr15] has the same scope as the CAVITY package: to provide the necessary means for reliable gyrotron cavity designs. Its two main components are COLDCAV and EVRIDIKI, in a similar fashion as COLDC and SELFT are central to CAVITY; other routines are ISTART (for starting current calculations), BEAMSHIFT, INTER and SELF (self-consistent solution of gyrotron interaction). EURIDICE is written in Fortran and parallelized with MPI.

COLDCAV

Coldcav finds the cold-cavity profiles of given modes in a defined cavity, and returns the corresponding eigenfrequencies and quality factors. It also calculates the necessary z-dependent eigenvalues in advance (what is SCNCHI and CHIMP in CAVITY). COLDCAV can consider broadband boundary conditions as described in [WATJ15].

EVRIDIKI

EVRIDIKI simulates self-consistent time-dependent multimode interactions in gyrotron cavities. Apart from the basic physical (such magnetic field strength, beam radius, voltage, current and pitch factor) and numerical parameters (such as time step, number of electron beamlets and phases), several advanced parameters can be set, such as spreads in γ, g and r_b as well as parabolic magnetic field taper. The electron beam can also be read-in particle for particle, thus providing arbitrarily shaped beams. The code has originally been designed for simulations of hollow cavities and of co-axial cavities with aligned corrugated insert, but interfaces exist to simulate other geometries such as cavities with misaligned insert (e.g. SCNCHIMP output) or with inner-outer corrugation.

C.2.3 Design-o-mat (Franck 2012-2016)

As follows from the considerations in Section 2.1.2, the gyrotron design outcome depends on the particular design process, i.e. which parameters are calculated in which order respecting which boundaries. For the gyrotron design in this work, MAXPO usually returned values that were much more pessimistic than the outcome of corresponding interaction and MIG simulations, and the code does not foresee that the user can change the order of parameter calculation. Therefore, a completely new C++ routine, named Design-o-mat, was written to serve the same purpose as MAXPO. Its internal calculation scheme is given in Fig. 2.1.

As of 2016, Design-o-mat can be used in three operation schemes. In the standard scheme, the user provides an input file with frequency and output power as the key parameters of a gyrotron design. Within the technical boundary conditions, the code then suggests optimum parameters (such as operating mode, cavity length, beam voltage and current) which may or may not be respected by the user during the next input. The code also returns interesting values such as main competitors, time scales and parameters for launcher simulations. After the design for the main mode is complete, the code calculates the corresponding values for all other modes in a multi-frequency series.

The other two operation schemes focus on multi-frequency design, based on the considerations described in Section 2.1.3. The more user-oriented scheme displays all mode series with central mode in a given range of azimuthal index, caustic radius, eigenvalue and frequency, and the user can analyze their suitability for multi-frequency operation in more detail. The other scheme just displays the most suitable series.

C.2.4 SCNCHIMP (Franck 2014-2016)

This code had originally been developed to fix the radial index issue in SCNCHI on the cost of speed and memory efficiency (which are much less an issue than 20 years ago). Since the coupling spectrum in SCNCHI in fact depends on the axially varying eigenvalue calculated later with CHIMP, it had been considered useful to combine both routines into one, hence the name SCNCHIMP. Some less important features of SCNCHI and CHIMP have been left out of the new code, but new features were added, most importantly the possibility to calculate field distributions in misaligned cavities based on the theory and methods described in Section 3.3.3, but also to include a magnetic field with parabolic taper (compatible to the EVRIDIKI input) and a tilted insert. SCNCHIMP has been tested for eigenvalues up to 180, azimuthal indices up to 60, misalignment up to 1 mm and root precision up to 10^{-9}. Modes with longitudinally varying dominant azimuthal index cannot be treated properly by the code; hoewever, those modes only occur for physically unacceptable large misalignment. SCNCHIMP is written in Fortran-95 and almost the entire code could be parallelized with OpenMP. See Section 3.3.

C.2.5 WickedQueen (Franck 2015-2016)

The WickedQueen C++ code has been developed in the context of this work to calculate voltage depression on beam electrons in misaligned cavities. Its fast and elegant calculations are based solely on the method of image charges and can therefore also be used to benchmark e.g. finite-element approaches. The output data of WickedQueen – the voltage depression distribution of the electrons – can be used as energy spread input data for interaction simulations, such as EVRIDIKI. More details are provided in [FAG+16a].

For an overview of the interconnection of the cavity design routines described in Sections C.2.1-C.2.5, see Fig. 2.5.

C.2.6 ARIADNE (Pagonakis, Xidakis, Vomvoridis 2007-2016)

ARIADNE is an electron beam optics code written in C++ for self-consistent calculation of electron trajectories in electrostatic, magnetostatic, and RF electromagnetic fields. It uses the finite-element method on a curvilinear mesh with second-order interpolation for the fields at electron positions. A multitude of commands from relatively low level to high level can be used to define physical components and phenomena typical for gyrotron MIGs and collectors such as metallic surfaces with defined potentials, emitters, magnet coils, beam neutralization, or secondary electrons.

C.2.7 Opic (Illy, 1999-2016)

Opic is a code package capable of simulating gyrotrons MIGs and collectors. Its core routine is ESRAY, a trajectory code written in C++ and parallelized with OpenMP. It uses a 2D non-orthogonal, but structured quadrilateral mesh and the finite-difference method to solve the Poisson equation. Further routines are dedicated to grid generation (GRIDGEN), magnetic field generation from coil definitions (MAGGEN) and visualization/data output (OVIS). As of 2016, a unique feature to ESRAY is its capability to include effects from emitter surface inhomogeneities auch as roughness and varying temperature.

Since ARIADNE and ESRAY use complementary numerical approaches, the codes can be – and have been – used to verify each other.

List of Figures

List of Tables

Bibliography

[ABG$^+$15] S. Alberti, F. Braunmueller, J. Genoud, J.-P. Hogge, T. M. Tran, M. Q. Tran, K. Avramidis, I. Gr. Pagonakis, J. Jin, S. Illy, G. Gantenbein, J. Jelonnek, and F. Cismondi. Dual-frequency, 126/84 GHz, 1 MW gyrotron for the upgrade of the TCV EC-system (talk W2E-4). In *40th International Conference on Infrared, Millimeter, and Terahertz Waves (IRMMW-THz)*, Aug 2015.

[AD02] M.I. Airila and O. Dumbrajs. Spatio-temporal chaos in the transverse section of gyrotron resonators. *IEEE Transactions on Plasma Science*, 30(3):846–850, June 2002.

[ADK$^+$08] K. A. Avramidis, O. Dumbrajs, S. Kern, I. Gr. Pagonakis, and J. L. Vomvoridis. Mode selection for a 170 GHz, 1 MW gyrotron (talk P4.105). *Europhysics Conference Abstracts (ECA)*, 32D, 2008. Proceedings of the 35th EPS Conference on Plasma Physics, 9-13 June 2008.

[AM07] W. R. Alexander and L. E. McKinley, editors. *Deep Geological Disposal of Radioactive Waste*. Radioactivity in the Environment. Elsevier Ltd., Oxford, 1st edition, 2007.

[APIV12] K. A. Avramidis, I. Gr. Pagonakis, C. T. Iatrou, and J. L. Vomvoridis. EURIDICE: A code-package for gyrotron interaction simulations and cavity design. *EPJ Web of Conferences*, 32:04016, 2012.

[AS64] Milton Abramowitz and Irene A. Stegun, editors. *Handbook of Mathematical Functions*. Dover Publications, Inc., New York, 9th Dover, 10th GPO printing edition, 1964.

[Avr06] Konstantinos A. Avramidis. *Design and Simulation of Coaxial Gyrotrons (with Emphasis on Second Harmonics) (in Greek)*. PhD thesis, NTU Athens, 2006.

[Avr15] K. A. Avramidis. Investigations and advanced concepts on gyrotron interaction modeling and simulations. *Physics of Plasmas*, 22(12), 2015.

[BBM$^+$05] J. D. Baena, J. Bonache, F. Martin, R. M. Sillero, F. Falcone, T. Lopetegi, M. A. G. Laso, J. Garcia-Garcia, I. Gil, M. F. Portillo, and M. Sorolla. Equivalent-circuit models for split-ring resonators and complementary split-ring resonators coupled to planar transmission lines. *IEEE Transactions on Microwave Theory and Techniques*, 53(4):1451–1461, April 2005.

[BC92] J. J. Barroso and R. A. Correa. Design of a TE42,7 coaxial cavity for a 1 MW, 280 GHz gyrotron. *International Journal of Infrared and Millimeter Waves*, 13(4):443–455, 1992.

[BCD$^+$84] G. F. Brand, Z. Chen, N. G. Douglas, M. Gross, J. Y. L. Ma, and L. C. Robinson. A tunable millimetre-submillimetre gyrotron. *International Journal of Electronics*, 57(6):863–870, 1984.

[BCM89] J. J. Barroso, R. A. Correa, and A. Montes. Optimization of RF field profiles for enhanced gyrotron efficiency. *International Journal of Electronics*, 67(3):457–484, 1989.

[BDKT97] O. Braz, G. Dammertz, M. Kuntze, and M. Thumm. D-band frequency step-tuning of a 1 MW gyrotron using a Brewster output window. *International Journal of Infrared and Millimeter Waves*, 18(8):1465–1477, 1997.

[BDN14] V. L. Bakunin, G. G. Denisov, and Yu. V. Novozhilova. Frequency and phase stabilization of a multimode gyrotron with megawatt power by an external signal. *Technical Physics Letters*, 40(5):382–385, 2014.

[Ber11] Matthias H. Beringer. *Design studies towards a 4 MW 170 GHz coaxial-cavity gyrotron*. PhD thesis, KIT, Karlsruhe, 2011.

[BIJ+09] M. H. Beringer, S. Illy, J. Jin, S. Kern, J. C. Rode, and M. Thumm. Further design steps towards a 4 MW 170 GHz coaxial-cavity gyrotron (talk T4C04.0154). In *34th International Conference on Infrared, Millimeter, and Terahertz Waves (IRMMW-THz)*, Sept 2009.

[BL86] J. M. Baird and W. Lawson. Magnetron injection gun (MIG) design for gyrotron applications. *International Journal of Electronics*, 61(6):953–967, 1986.

[BNP76] V. L. Bratman, S. L. Novozhilov, and M. I. Petelin. Frequency tuning in a gyromonotron with a microwave circuit in the form of a conical waveguide. *Elektronnaya Tekhnika*, 11:46–49, 1976.

[Bor91] E. Borie. *Review of Gyrotron Theory*. KfK Scientific reports 4898. Kernforschungszentrum Karlsruhe, Karlsruhe, 1991.

[Bre91] Jork Bretting, editor. *Technische Röhren : Grundlagen, Funktionen, Anwendung*. Hüthig, Heidelberg, 1991.

[Bur01] A. V. Burov. Busch's theorem for mappings. In *19th Particle Accelerator Conference (PAC 2001)*, pages 1708–1710, 2001.

[Bus26] H. Busch. Berechnung der Bahn von Kathodenstrahlen im axialsymmetrischen elektromagnetischen Felde. *Annalen der Physik*, 386(25):974–993, 1926.

[CAB+13] F. Cismondi, F. Albajar, T. Bonicelli, EGYC, and Thales. EU development program for the 1 MW gyrotron for ITER (Talk T4 E-16.15). In *38th International Conference on Infrared, Millimeter, and Terahertz Waves (IRMMW-THz)*, Sept 2013.

[CB93] R. A. Correa and J. J. Barroso. Space charge effects of gyrotron electron beams in coaxial cavities. *International Journal of Electronics*, 74(1):131–136, 1993.

[CD84] K. R. Chu and D. Dialetis. Theory of harmonic gyrotron oscillator with slotted resonant structure. *International Journal of Infrared and Millimeter Waves*, 5(1):37–56, 1984.

[Cho14] Amitavo Roy Choudhury. *Investigations of After Cavity Interaction in Gyrotrons Including the Effect of Non-uniform Magnetic Field.* PhD thesis, KIT, Karlsruhe, 2014.

[CSB+13] P. Ciais, C. Sabine, G. Bala, L. Bopp, V. Brovkin, J. Canadell, A. Chhabra, R. DeFries, J. Galloway, M. Heimann, C. Jones, C. Le Quéré, R.B. Myneni, S. Piao, and P. Thornton. *Carbon and other Biogeochemical Cycles.* Climate Change 2013: The Physical Science Basis. Contributions of Working Group I to the Fifth Assessment Report of the Intergovernmental Panel on Climate Change [T.F. Stocker, D. Qin, G.-K. Plattner, M. Tignor, S.K. Allen, J. Boschung, A. Nauels, Y. Xia, V. Bex and P.M. Midgley (eds.)]. Cambridge University Press, Cambridge, UK and New York, NY, USA, 2013.

[DAFJ14] O. Dumbrajs, K. A. Avramidis, J. Franck, and J. Jelonnek. On the dependence of the efficiency of a 240 GHz high-power gyrotron on the displacement of the electron beam and on the azimuthal index. *Physics of Plasmas*, 21(1), 2014.

[Dat13] NEWMDB: IAEA's Net-Enabled Radioactive Waste Management Database. http://newmdb.iaea.org/about.aspx, 2013. Accessed 2015-09-20.

[DII+03] O. Dumbrajs, T. Idehara, Y. Iwata, S. Mitsudo, I. Ogawa, and B. Piosczyk. Hysteresis-like effects in gyrotron oscillators. *Physics of Plasmas*, 10(5):1183–1186, 2003.

[DK81] A. T. Drobot and K. Kim. Space charge effects on the equilibrium of guided electron flow with gyromotion. *International Journal of Electronics*, 51(4):351–367, 1981.

[DK95] O. Dumbrajs and J. Koponen. Magnetic field tapering in the kfk coaxial gyrotron. *International Journal of Infrared and Millimeter Waves*, 16(3):473–482, 1995.

[DKM+92] G. G. Denisov, A. N. Kuftin, V. I. Malygin, N. P. Venediktov, D. V. Vinogradov, and V. E. Zapevalov. 110 GHz gyrotron with a built-in high-efficiency converter. *International Journal of Electronics*, 72(5-6):1079–1091, 1992.

[DKR05] O. Dumbrajs, H. Kalis, and A. Reinfelds. Are coaxial super power gyrotrons feasible? *International Journal of Infrared and Millimeter Waves*, 26(6):787–805, 2005.

[DM98] O. Dumbrajs and A. Möbius. Tunable coaxial gyrotron for plasma heating and diagnostics. *International Journal of Electronics*, 84(4):411–419, 1998.

[DN04] O. Dumbrajs and G. S. Nusinovich. Coaxial gyrotrons: past, present, and future (review). *IEEE Transactions on Plasma Science*, 32(3):934–946, June 2004.

[DP90] O. Dumbrajs and B. Piosczyk. Resonator for a frequency-step tunable gyrotron. *International Journal of Electronics*, 68(5):885–890, 1990.

[DP95] O. Dumbrajs and A.B. Pavelyev. Symmetry breaking in coaxial cavities and its influence on gyrotron operation. In K. Herschbach, W. Maurer, and J. E. Vetter, editors, *Fusion Technology 1994 - Proceedings of the 18th Symposium on Fusion Technology, Karlsruhe, Germany*, volume 1, pages 521 – 524. Elsevier, Oxford, 1995.

[DZ04] O. Dumbrajs and G. I. Zaginaylov. Ohmic losses in coaxial gyrotron cavities with corrugated insert. *IEEE Transactions on Plasma Science*, 32(3):861–866, June 2004.

[EC09] L. A. El-Guebaly and L. C. Cadwallader. Recent developments in safety and environmental aspects of fusion experiments and power plants. In A. Aasen and P. Olsson, editors, *Nuclear Reactors, Nuclear Fusion and Fusion Engineering*, pages 321–365. Nova Science Publ., Inc., 2009.

[Edg93] Christopher J. Edgcombe, editor. *Gyrotron Oscillators: Their Principles and Practice*. Taylor & Francis, Washington, DC, 1993.

[EG06] L. A. El-Guebaly. Managing fusion high-level waste – a strategy for burning the long-lived products in fusion devices. *Fusion Engineering and Design*, 81(8-14):1321 – 1326, 2006. 7th International Symposium on Fusion Nuclear Technology (ISFNT-7).

[EH89] Joseph Eichmeier and Hinrich Heynisch, editors. *Handbuch der Vakuumelektronik*. Oldenbourg, München [u.a.], 1989.

[EUR11] EUROfusion. https://www.euro-fusion.org/2011/09/progress-in-fusion, 2011.

[FAG⁺15] J. Franck, K. A. Avramidis, G. Gantenbein, S. Illy, J. Jin, M. Thumm, and J. Jelonnek. A generic mode selection strategy for high-order mode gyrotrons operating at multiple frequencies (featured article). *Nuclear Fusion*, 55(1):013005, 2015.

[FAG+16a] J. Franck, K. A. Avramidis, G. Gantenbein, S. Illy, I. G. Pagonakis, M. Thumm, and J. Jelonnek. Direct voltage depression calculation of arbitrary electron beams in misaligned coaxial gyrotron cavities. *IEEE Transactions on Electron Devices*, 63(9):3740–3746, Sept 2016.

[FAG+16b] J. Franck, K. A. Avramidis, G. Gantenbein, S. Illy, I. Gr. Pagonakis, M. Thumm, and J. Jelonnek. Insert misalignment in coaxial gyrotrons: physical effects and numerical treatment (unpublished presentation). In *5th ITG International Vacuum Electronics Workshop*, September 2016.

[FAI13] J. Franck, K. Avramidis, and S. Illy. Mode selection for 240-300 GHz gyrotrons (unpublished presentation). In *25th Joint Russian-German Workshop on ECRH and Gyrotrons*, June 2013.

[FAI+14] J. Franck, K. A. Avramidis, S. Illy, J. Jelonnek, I. Gr. Pagonakis, and M. Thumm. Design aspects for DEMO-compatible 2 MW gyrotrons: Electron gun and cavity (unpublished presentation). In *4th ITG International Vacuum Electronics Workshop*, October 2014.

[FAP+15] J. Franck, K. A. Avramidis, I. G. Pagonakis, S. Illy, G. Gantenbein, M. Thumm, and J. Jelonnek. Multi-frequency design of a 2 MW coaxial-cavity gyrotron for DEMO. In *40th International Conference on Infrared, Millimeter, and Terahertz waves (IRMMW-THz)*, Aug 2015.

[FGPY77] V. A. Flyagin, A. V. Gaponov, M. I. Petelin, and V. K. Yulpatov. The gyrotron. *IEEE Transactions on Microwave Theory and Techniques*, 25(6):514–521, June 1977.

[FHKS89] B. Fischer, L. Hahn, C. Küppers, and M. Sailer. *Der Atommüll-Report – "Entsorgung", Wiederaufarbeitung, Lagerung: das offene Ende der Atomwirtschaft*. Rasch u. Röhring, Hamburg, 1989.

[FIA+14a] J. Franck, S. Illy, K. A. Avramidis, J. Jelonnek, and M. Thumm. Mode selection and resonator design for DEMO gyrotrons. In *15th IEEE International Vacuum Electronics Conference (IVEC)*, pages 31–32, April 2014.

[FIA+14b] J. Franck, S. Illy, K. A. Avramidis, J. Jelonnek, and M. Thumm. Mode selection and resonator design for DEMO gyrotrons (unpublished presentation). In *4th KIT Fusion PhD Seminar*, June 2014.

[FKM+94] V. A. Flyagin, V. I. Khizhnyak, V. N. Manuilov, A. B. Pavelyev, V. G. Pavelyev, B. Piosczyk, G. Dammertz, O. Höchtl, C. T. Iatrou, S. Kern, H.-U. Nickel, M. Thumm, A. Wien, and O. Dumbrajs. Development of a 1.5-MW coaxial gyrotron at 140 GHz. In *19th International Conference on Infrared and Millimeter Waves*, pages 75–76, 1994.

[FKW+14] G. Federici, R. Kemp, D. Ward, C. Bachmann, T. Franke, S. Gonzalez, C. Lowry, M. Gadomska, J. Harman, B. Meszaros, C. Morlock, F. Romanelli, and R. Wenninger. Overview of EU DEMO design and R&D activities. *Fusion Engineering and Design*, 89(7-8):882 – 889, 2014. 11th International Symposium on Fusion Nuclear Technology-11 (ISFNT-11).

[Fla12] Jens H. Flamm. *Diffraction and scattering in launchers of quasi-optical mode converters for gyrotrons*. PhD thesis, KIT, Karlsruhe, 2012.

[FLS62] Richard P. Feynman, Robert B. Leighton, and Matthew Sands. *Mainly Electromagnetism and Matter*. The Feynman Lectures on Physics II. California Institute of Technology, Pasadena, CA, USA, 1962.

[FPA+15a] J. Franck, I. Pagonakis, K. Avramidis, G. Gantenbein, S. Illy, M. Thumm, and J. Jelonnek. Multi-frequency behavior of a 2 MW coaxial-cavity DEMO gyrotron (unpublished presentation). In *27th Joint Russian-German Workshop on ECRH and Gyrotrons*, July 2015.

[FPA+15b] J. Franck, I. Pagonakis, K. Avramidis, G. Gantenbein, S. Illy, M. Thumm, and J. Jelonnek. Multi-frequency behavior of a 2 MW coaxial-cavity gyrotron for DEMO (unpublished presentation). In *5th KIT Fusion PhD Seminar*, June 2015.

[FPA+15c] J. Franck, I. Gr. Pagonakis, K. A. Avramidis, G. Gantenbein, S. Illy, M. Thumm, and J. Jelonnek. Magnetron injection gun for a 238 GHz 2 MW coaxial-cavity gyrotron. In *German Microwave Conference (GeMiC)*, pages 260–263, March 2015.

[Fra13a] J. Franck. Gyrotrons for DEMO: mode selection strategies (unpublished poster). In *Wissenschaftliche Beiträge des KIT zur 2. Jahrestagung des KIT-Zentrums Energie – Doktorandensymposium*, June 2013.

[Fra13b] J. Franck. Mode selection for a 240 GHz coaxial gyrotron (unpublished presentation). In *3rd KIT Fusion PhD Seminar*, June 2013.

[FYYL09] W. Fu, Y. Yan, X. Yuan, and S. Liu. Two-beam magnetron injection guns for coaxial gyrotron with two electron beams. *Physics of Plasmas*, 16(2), 2009.

[GB90] G. Gantenbein and E. Borie. Gyrotron with a tapered external magnetic field. *International Journal of Infrared and Millimeter Waves*, 11(7):837–850, 1990.

[GBB$^+$15] S. Garavaglia, W. Bin, A. Bruschi, G. Granucci, G. Grossetti, J. Jelonnek, A. Moro, N. Rispoli, D. Strauss, Q. M. Tran, and T. Franke. Preliminary conceptual design of DEMO EC system. *AIP Conference Proceedings*, 1689(1):5, 2015.

[GC84] A. K. Ganguly and K. R. Chu. Limiting current in gyrotrons. *International Journal of Infrared and Millimeter Waves*, 5(1):103–121, 1984.

[GCD$^+$06] A. Godeke, D. Cheng, D. R. Dietderich, P. Ferracin, S. O. Prestemon, G. Sabbi, and R. M. Scanlan. Limits of NbTi and Nb3Sn, and development of W&R Bi-2212 high field accelerator magnets, Sep 2006.

[GDZ$^+$96] A. L. Goldenberg, G. G. Denisov, V. E. Zapevalov, A. G. Litvak, and V. A. Flyagin. Cyclotron resonance masers: State of the art. *Radiophysics and Quantum Electronics*, 39(6):423–446, 1996.

[GFG$^+$81] A. V. Gaponov, V. A. Flyagin, A. L. Gol'denberg, G. S. Nusinovich, Sh. E. Tsimring, V. G. Usov, and S. N. Vlasov. Powerful millimetre-wave gyrotrons (invited paper). *International Journal of Electronics*, 51(4):277–302, 1981.

[GGG$^+$12] M. Yu. Glyavin, N. S. Ginzburg, A. L. Gol'denberg, G. G. Denisov, A. G. Luchinin, V. N. Manuilov, V. E. Zapevalov, and I. V. Zotova. THz gyrotrons: Status and possible optimizations. *Terahertz Science and Technology*, 5(2):67–77, 2012.

[GGM$^+$16] N. S. Ginzburg, M. Y. Glyavin, A. M. Malkin, V. N. Manuilov, R. M. Rozental, A. S. Sedov, A. S. Sergeev, V. Y. Zaslavsky,

I. V. Zotova, and T. Idehara. Improvement of stability of high cyclotron harmonic operation in the double-beam thz gyrotrons. *IEEE Transactions on Plasma Science*, 44(8):1303–1309, Aug 2016.

[GGR+15] G. Granucci, S. Garavaglia, D. Ricci, G. Artaserse, F. Belli, W. Bin, G. Calabr Cavinato, D. Farina, L. Figini, A. Moro, G. Ramogida, C. Sozzi, O. Tudisco, and FTU team. Experiments and modeling on FTU tokamak for EC assisted plasma start-up studies in ITER-like configuration. *Nuclear Fusion*, 55(9):093025, 2015.

[Gil86] Alexander S. Gilmour, Jr. *Microwave Tubes*. Artech House, Inc., Dedham, MA, 1986.

[GLMK08] M. Yu. Glyavin, A. G. Luchinin, M. V. Morozkin, and V. I. Khizhnyak. Smooth wideband tuning of the operating frequency of a gyrotron. *Radiophysics and Quantum Electronics*, 51(1):57–63, 2008.

[Goo14] Jeremy Good. Eine Lösung für das Heliumproblem. *Physik in unserer Zeit*, 45(5):238–242, 2014.

[GPY67] A. V. Gaponov, M. I. Petelin, and V. K. Yulpatov. The induced radiation of excited classical oscillators and its use in high-frequency electronics. *Radiophysics and Quantum Electronics*, 10(9-10):794–813, 1967.

[GS08] Peter Gruss and Ferdi Schüth, editors. *Die Zukunft der Energie*. Beck, München, 2008.

[HBG+05] M. K. Hornstein, V. S. Bajaj, R. G. Griffin, K. E. Kreischer, I. Mastovsky, M. A. Shapiro, J. R. Sirigiri, and R. J. Temkin. Second harmonic operation at 460 GHz and broadband continuous frequency tuning of a gyrotron oscillator. *IEEE Transactions on Electron Devices*, 52(5):798–807, May 2005.

[HCC+14] O. A. Hurricane, D. A. Callahan, D. T. Casey, P. M. Celliers, C. Cerjan, E. L. Dewald, T. R. Dittrich, T. Döppner, D. E. Hinkel, L. F. Berzak Hopkins, J. L. Kline, S. Le Pape, T. Ma, A. G. MacPhee, J. L. Milovich, A. Pak, H.-S. Park, P. K. Patel, B. A. Remington, J. D. Salmonson, P. T. Springer, and R. Tommasini. Fuel gain exceeding unity in an inertially confined fusion implosion. *Nature*, 506:343–348, Feb. 2014.

[HG77] J. L. Hirshfield and V. L. Granatstein. The electron cyclotron maser - an historical survey. *IEEE Transactions on Microwave Theory and Techniques*, 25(6):522–527, Jun 1977.

[HKO+08] R. Hirose, T. Kamikado, Y. Okui, H. Miyata, K. Shibutani, O. Ozaki, and K. Sakamoto. Development of 7 T cryogen-free superconducting magnet for gyrotron. *IEEE Transactions on Applied Superconductivity*, 18(2):920–923, June 2008.

[HMHH05] M. Hamada, T. Miyatake, S. Hayashi, and R. Hirose. Development in superconducting wires and magnets. In *Kobelco Technology Review*, pages 103–109. KOBELCO Kobe Steel, Ltd., 26 edition, 2005.

[HYL15] Shenyong Hou, Sheng Yu, and Hongfu Li. Research on a 170 GHz, 2 MW coaxial cavity gyrotron with inner-outer corrugation. *Physics of Plasmas*, 22(3), 2015.

[IALT11] Z. C. Ioannidis, K. A. Avramides, G. P. Latsas, and I. G. Tigelis. Azimuthal mode coupling in coaxial waveguides and cavities with longitudinally corrugated insert. *IEEE Transactions on Plasma Science*, 39(5):1213–1221, May 2011.

[IDT08] Z. C. Ioannidis, O. Dumbrajs, and I. G. Tigelis. Linear and non-linear inserts for genuinely wideband continuous frequency tunable coaxial gyrotron cavities. *International Journal of Infrared and Millimeter Waves*, 29(4):416–423, 2008.

[IKP96] C. T. Iatrou, S. Kern, and A. B. Pavelyev. Coaxial cavities with corrugated inner conductor for gyrotrons. *IEEE Transactions on Microwave Theory and Techniques*, 44(1):56–64, Jan. 1996.

[IL09] Harald Ibach and Hans Lüth, editors. *Festkörperphysik: Einführung in die Grundlagen.* Springer-Lehrbuch. Springer, Berlin, Heidelberg, 7th edition, 2009.

[IPP] IPP. https://www.ipp.mpg.de/93212/17.jpg. Accessed 2016-08-22.

[IPP03] IPP. *Kernfusion – Berichte aus der Forschung*, volume 2. Max-Planck-Institut für Plasmaphysik (IPP), Garching bei München, 2003.

[IZJ15] S. Illy, J. Zhang, and J. Jelonnek. Gyrotron electron gun and collector simulation with the ESRAY beam optics code (poster P1.1). In *16th IEEE International Vacuum Electronics Conference (IVEC)*, April 2015.

[Jac99] John D. Jackson. *Classical Electrodynamics.* Wiley, New York, 3rd edition, 1999.

[JGTJ15] J. Jin, G. Gantenbein, M. Thumm, and J. Jelonnek. A hybrid-type 170 GHz gyrotron launcher for the TE32,9 mode. In *16th IEEE International Vacuum Electronics Conference (IVEC)*, April 2015.

[Jin13] Jianbo Jin. Private communication, 2013.

[JK15] Jianbo Jin and Parth Kalaria. Private communication, 2015.

[KAF+15a] P. Kalaria, K. A. Avramidis, J. Franck, G. Gantenbein, S. Illy, I. G. Pagonakis, M. Thumm, and J. Jelonnek. Interaction circuit design and RF behavior of a 236 GHz gyrotron for DEMO. In *German Microwave Conference (GeMiC)*, pages 264–267, March 2015.

[KAF+15b] P. C. Kalaria, K. A. Avramidis, J. Franck, S. Illy, I. G. Pagonakis, M. Thumm, and J. Jelonnek. Multi-frequency operation of DEMO gyrotron with realistic electron beam parameters. In *16th IEEE International Vacuum Electronics Conference (IVEC)*, April 2015.

[KAF+16a] P. Kalaria, K. A. Avramidis, J. Franck, G. Gantenbein, S. Illy, I. G. Pagonakis, M. Thumm, and J. Jelonnek. A fast frequency step-tunable 236 GHz gyrotron design for DEMO. In *German Microwave Conference (GeMiC)*, pages 377–380, March 2016.

[KAF+16b] P. C. Kalaria, K. A. Avramidis, J. Franck, G. Gantenbein, S. Illy, I. G. Pagonakis, M. Thumm, and J. Jelonnek. Investigation on mode eigenvalue limits for stable 236 GHz, 1 MW-class gyrotron operation. In *17th IEEE International Vacuum Electronics Conference (IVEC)*, April 2016.

[KAF+16c] P. C. Kalaria, K. A. Avramidis, J. Franck, G. Gantenbein, S. Illy, I. Gr. Pagonakis, M. Thumm, and J. Jelonnek. Systematic cavity design approach for a multi-frequency gyrotron for DEMO and study of its RF behavior. *Physics of Plasmas*, 23(9), 2016.

[KAF+17] P. C. Kalaria, K. A. Avramidis, J. Franck, G. Gantenbein, S. Illy, J. Jin, I. Gr. Pagonakis, M. Thumm, and J. Jelonnek. RF behavior and launcher design for a fast frequency step-tunable 236 GHz gyrotron for DEMO. *Frequenz – Journal of RF-Engineering and Telecommunications*, 71(3-4), 2017.

[KBT04] Machavaram V. Kartikeyan, Edith Borie, and Manfred K. A. Thumm. *Gyrotrons*. Advanced texts in physics. Springer, Berlin, 2004.

[KDK+00] K. Koppenburg, G. Dammertz, M. Kuntze, B. Piosczyk, and M. Thumm. Fast-frequency step-tunable high-power gyrotron with hybrid magnet system. In *IEEE International Vacuum Electronics Conference (IVEC)*, page 2, May 2000.

[KDK+01] K. Koppenburg, G. Dammertz, M. Kuntze, B. Piosczyk, and M. Thumm. Fast frequency-step-tunable high-power gyrotron with hybrid-magnet-system. *IEEE Transactions on Electron Devices*, 48(1):101–107, Jan. 2001.

[KDW+84] K. E. Kreischer, B. G. Danly, P. Woskoboinikow, W. J. Mulligan, and R. J. Temkin. Frequency pulling and bandwidth measurements of a 140 GHz pulsed gyrotron. *International Journal of Electronics*, 57(6):851–862, 1984.

[Ker96] Stefan Kern. *Numerische Simulation der Gyrotron-Wechselwirkung in koaxialen Resonatoren*. PhD thesis, FZKA, Karlsruhe, 1996.

[KGOP90] Ronald L. Klueh, D. S. Gelles, M. Okada, and N. H. Packan. Reduced activation materials for fusion reactors, 1990.

[KIK+15] K. Kim, K. Im, H.C. Kim, S. Oh, J.S. Park, S. Kwon, Y.S. Lee, J.H. Yeom, C. Lee, G-S. Lee, G. Neilson, C. Kessel, T. Brown, P. Titus, D. Mikkelsen, and Y. Zhai. Design concept of K-DEMO for near-term implementation. *Nuclear Fusion*, 55(5):053027, 2015.

[KJT13] M. V. Kartikeyan, J. Jelonnek, and M. Thumm. A 1.0 – 1.3 MW CW, 238 GHz conventional cavity gyrotron (poster Th P3-14). In *38th International Conference on Infrared, Millimeter, and Terahertz Waves (IRMMW-THz)*, Sept 2013.

[KM69] Werner Kleen and Rudolf Müller, editors. *Laser*. Springer, Berlin, 1969.

[Kob15] Thorsten Kobarg. Private communication, 2015.

[KSM+04] A Kasugai, K Sakamoto, R Minami, K Takahashi, and T Imai. Study of millimeter wave high-power gyrotron for long pulse operation. *Nuclear Instruments and Methods in Physics Research Section A: Accelerators, Spectrometers, Detectors and Associated Equipment*, 528(1-2):110 – 114, 2004. Proceedings of the 25th International Free Electron Laser Conference, and the 10th FEL Users Workshop.

[KSN00] M Kikuchi, Y Seki, and K Nakagawa. The advanced SSTR. *Fusion Engineering and Design*, 48(3-4):265 – 270, 2000.

[LB82] Y. Y. Lau and L. R. Barnett. A low magnetic field gyrotron – gyro-magnetron. *International Journal of Electronics*, 53(6):693–698, 1982.

[LCV+05] Y. Lechon, H. Cabal, M. Varela, R. Saez, C. Eherer, M. Baumann, J. Düweke, T. Hamacher, and G.C. Tosato. A global energy model with fusion. *Fusion Engineering and Design*, 75-79(0):1141 – 1144, 2005. 23rd Symposium of Fusion Technology (SOFT 23).

[LDM+12] A. G. Litvak, G. G. Denisov, V. E. Myasnikov, E. M. Tai, E. V. Sokolov, and V. I. Ilin. Recent development results in Russia of megawatt power gyrotrons for plasma fusion installations. *EPJ Web of Conferences*, 32:04003, 2012.

[LL97] Lew D. Landau and Ewgenij M. Lifschitz. *Lehrbuch der theoretischen Physik*, volume II: Klassische Feldtheorie. Verlag Harry Deutsch, Thun & Frankfurt/Main, 12th edition, 1997.

[LMK+95] V. K. Lygin, V. N. Manuilov, A. N. Kuftin, A. B. Pavelyev, and B. Piosczyk. Inverse magnetron injection gun for a coaxial 1.5 MW, 140 GHz gyrotron. *International Journal of Electronics*, 79(2):227–235, 1995.

[LYF+07] S. Liu, X. Yuan, W. Fu, Y. Yan, Y. Zhang, H. Li, and R. Zhong. The coaxial gyrotron with two electron beams. I. linear theory and nonlinear theory. *Physics of Plasmas*, 14(10), 2007.

[Mei66] Hans H. Meinke. *Einführung in die Elektrotechnik höherer Frequenzen*. Springer, Berlin, 1966.

[MIP+12] A. Malygin, S. Illy, I. Pagonakis, B. Piosczyk, S. Kern, J. Weggen, M. Thumm, K. Avramides, L. Ives, and D. Marsden. Design of a 10 kW/28 GHz gyrotron with a segmented emitter using controlled-porosity reservoir cathodes. In *13th IEEE International Vacuum Electronics Conference (IVEC)*, pages 331–332, April 2012.

[MK94] Theo Mayer-Kuckuk. *Kernphysik*. Teubner-Studienbücher Physik. Teubner, Stuttgart, 6th edition, 1994.

[MRR+16] M. M. Melnikova, A. G. Rozhnev, N. M. Ryskin, A. V. Tyshkun, M. Y. Glyavin, and Y. V. Novozhilova. Frequency stabilization of a 0.67-thz gyrotron by self-injection locking. *IEEE Transactions on Electron Devices*, 63(3):1288–1293, March 2016.

[NAB+06] F. Najmabadi, A. Abdou, L. Bromberg, T. Brown, V. C. Chan, M. C. Chu, F. Dahlgren, L. El-Guebaly, P. Heitzenroeder, D. Henderson, H. E. St. John, C. E. Kessel, L. L. Lao, G. R. Longhurst, S. Malang, T. K. Mau, B. J. Merrill, R. L. Miller, E. Mogahed, R. L. Moore, T. Petrie, D. A. Petti, P. Politzer, A. R. Raffray, D. Steiner, I. Sviatoslavsky, P. Synder, G. M. Syaebler, A. D. Turnbull, M. S. Tillack, L. M. Waganer, X. Wang, P. West, and P. Wilson. The ARIES-AT advanced tokamak, advanced technology fusion power plant. *Fusion Engineering and Design*, 80(1-4):3 – 23, 2006. Aries - AT Special Issue.

[NBGT11] E. A. Nanni, A. B. Barnes, R. G. Griffin, and R. J. Temkin. THz dynamic nuclear polarization NMR. *IEEE Transactions on Terahertz Science and Technology*, 1(1):145–163, Sept 2011.

[NPA+10] G. S. Nusinovich, R. Pu, T. M. Antonsen, O. V. Sinitsyn, J. Rodgers, V. L. Granatsten, A. Mohamed, J. Silverman, M. Al-Sheikhly, G. M. Milikh, and Y. S. Dimant. Development of THz gyrotrons with pulse solenoids for detecting concealed radioactive materials. In *35th International Conference on Infrared, Millimeter, and Terahertz Waves (IRMMW-THz)*, Sept 2010.

[NRI09] J. Neilson, M. Read, and L. Ives. Design of a permanent magnet gyrotron for active denial systems. In *10th IEEE International Vacuum Electronics Conference (IVEC)*, pages 92–93, April 2009.

[NTP14] G. S. Nusinovich, M. K. A. Thumm, and M. I. Petelin. The gyrotron at 50: Historical overview. *Journal of Infrared, Millimeter, and Terahertz Waves*, 35(4):325–381, 2014.

[Nus04] Gregory S. Nusinovich. *Introduction to the Physics of Gyrotrons*. John Hopkins Studies in Applied Physics. Johns Hopkins Univ. Press, Baltimore, 2004.

[O+14] K. A. Olive et al. Review of particle physics. *Chin. Phys.*, C38:090001, 2014.

[PBD+97] B. Piosczyk, O. Braz, G. Dammertz, C. T. Iatrou, S. Kern, M. Kuntze, A. Möbius, M. Thumm, V. A. Flyagin, V. I. Khishnyak, V. I. Malygin, A. B. Pavelyev, and V. E. Zapevalov. A 1.5-MW, 140-GHz, TE28,16-coaxial cavity gyrotron. *IEEE Transactions on Plasma Science*, 25(3):460–469, Jun 1997.

[PBD+98] B. Piosczyk, O. Braz, G. Dammertz, C.T. Iatrou, S. Illy, M. Kuntze, G. Michel, A. Möbius, M. Thumm, V.A. Flyagin,

V.I. Khishnyak, A.B. Pavelyev, and V.E. Zapevalov. Coaxial cavity gyrotron with dual RF beam output. *IEEE Transactions on Plasma Science*, 26(3):393–401, Jun 1998.

[PBD+99] B. Piosczyk, O. Braz, G. Dammertz, C.T. Iatrou, S. Illy, M. Kuntze, G. Michel, and M. Thumm. 165 GHz, 1.5 MW-coaxial cavity gyrotron with depressed collector. *IEEE Transactions on Plasma Science*, 27(2):484–489, April 1999.

[PDD+05] B. Piosczyk, G. Dammertz, O. Dumbrajs, S. Illy, J. Jin, W. Leonhardt, G. Michel, O. Prinz, T. Rzesnicki, M. Schmid, M. Thumm, and X. Yang. A 2 MW, 170 GHz coaxial cavity gyrotron - experimental verification of the design of main components. *Journal of Physics: Conference Series*, 25(1):24, 2005.

[PEK+12] L. Di Pace, L. A. El-Guebaly, B. Kolbasov, V. Massaut, and M. Zucchetti. Radioactive waste management of fusion power plants. In R. Abdel Rahman, editor, *Radioactive Waste*. InTech, 2012.

[PHA+08] I. G. Pagonakis, J. P. Hogge, S. Alberti, K. A. Avramides, and J. L. Vomvoridis. A new concept for the collection of an electron beam configured by an externally applied axial magnetic field. *IEEE Transactions on Plasma Science*, 36(2):469–480, April 2008.

[PHA+10] I. G. Pagonakis, J. P. Hogge, S. Alberti, B. Piosczyk, S. Illy, S. Kern, and C. Liévin. An additional criterion for gyrotron gun design. In *IEEE International Conference on Plasma Science*, June 2010.

[PHG+09] I. G. Pagonakis, J. P. Hogge, T. Goodman, S. Alberti, B. Piosczyk, S. Illy, T. Rzesnicki, S. Kern, and C. Liévin. Gun design criteria for the refurbishment of the first prototype of the EU 170GHz/2MW/CW coaxial cavity gyrotron for ITER. In *34th International Conference on Infrared, Millimeter, and Terahertz Waves (IRMMW-THz)*, Sept 2009.

[Pio01] B. Piosczyk. A novel 4.5-MW electron gun for a coaxial cavity gyrotron. *IEEE Transactions on Electron Devices*, 48(12):2938–2944, Dec. 2001.

[PPZ+16] I. Gr. Pagonakis, B. Piosczyk, J. Zhang, S. Illy, T. Rzesnicki, J.-P. Hogge, K. Avramidis, G. Gantenbein, M. Thumm, and J. Jelonnek. Electron trapping mechanisms in magnetron injection guns. *Physics of Plasmas*, 23(2), 2016.

[PR16] Ioannis Pagonakis and Sebastian Ruess. Private communication, 2016.

[PV04] I. Gr. Pagonakis and J. L. Vomvoridis. The self-consistent 3D trajectory code Ariadne for gyrotron beam tunnel simulation. In *29th International Conference on Infrared, Millimeter, and Terahertz Waves (IRMMW-THz)*, pages 657–658, 2004.

[PWI+16] I. G. Pagonakis, C. Wu, S. Illy, G. Gantenbein, and J. Jelonnek. Multistage depressed collector design approach for gyrotron. In *17th IEEE International Vacuum Electronics Conference (IVEC)*, April 2016.

[QLP+09] J. X. Qiu, B. Levush, J. Pasour, A. Katz, C. M. Armstrong, D. R. Whaley, J. Tucek, K. Kreischer, and D. Gallagher. Vacuum tube amplifiers. *IEEE Microwave Magazine*, 10(7):38–51, Dec 2009.

[QLY+14] M. Qin, Y. Luo, K. Yang, Y. Huang, H. Li, and S. Wang. Nonlinear theory of a corrugated coaxial-gyrotron with misaligned inner rod. *IEEE Transactions on Electron Devices*, 61(12):4247–4252, Dec. 2014.

[QYL+15] M. Qin, K. Yang, Y. Luo, Y. Huang, H. Li, and S. Wang. The study of a coaxial gyrotron with misaligned inner rod. *Vacuum*, 115(0):124 – 129, 2015.

[RBB+12] F. Romanelli, P. Barabaschi, D. Borba, G. Federici, L. Horton, R. Neu, D. Stork, and H. Zohm. A roadmap to the realization of fusion energy. *EFDA*, (12) 52/7.1.1.a, 2012.

[RBI+09] J. C. Rode, M. H. Beringer, S. Illy, S. Kern, B. Piosczyk, and M. Thumm. Design study of magnetron injection guns for a 4 MW 170 GHz coaxial gyrotron. In *10th IEEE International Vacuum Electronics Conference*, pages 96–97, April 2009.

[RPG+16] S. Ruess, I. G. Pagonakis, G. Gantenbein, S. Illy, T. Kobarg, T. Rzesnicki, M. Thumm, J. Weggen, and J. Jelonnek. An inverse magnetron injection gun for the KIT 2-MW coaxial-cavity gyrotron. *IEEE Transactions on Electron Devices*, 63(5):2104–2109, May 2016.

[RPK+10] T. Rzesnicki, B. Piosczyk, S. Kern, S. Illy, J. Jin, A. Samartsev, A. Schlaich, and M. Thumm. 2.2-MW record power of the 170-GHz European preprototype coaxial-cavity gyrotron for ITER. *IEEE Transactions on Plasma Science*, 38(6):1141–1149, June 2010.

[RRP+16] S. Ruess, T. Rzesnicki, I. G. Pagonakis, T. Kobarg, M. Fuchs, S. Illy, G. Gantenbein, A. M. Thumm, and J. Jelonnek. Experimental results and outlook of the 2 mw 170 ghz coaxial-cavity gyrotron towards long pulse operation. In *German Microwave Conference (GeMiC)*, pages 369–372, March 2016.

[RS83] Marshall N. Rosenbluth and Roald Z. Sagdeev. *Handbook of Plasma Physics*. Atomizdat / North-Holland Publishing Company, 1983.

[Rze07] Tomasz Rzesnicki. *Analyse eines neuartigen 1,5 MW, 170 GHz Prototyp-Gyrotrons mit koaxialem Resonator.* PhD thesis, FZKA, Karlsruhe, 2007.

[Sch15] Andreas Schlaich. *Time-dependent spectrum analysis of high power gyrotrons.* PhD thesis, KIT, Karlsruhe, 2015.

[SFK+06] M. I. Surin, A. S. Fix, V. E. Keilin, N. I. Kudriavsev [sic], S. M. Miklyaev, S. A. Shevchenko, and E. V. Sokolov. A prototype superconducting magnet for 170 GHz gyrotrons. *IEEE Transactions on Applied Superconductivity*, 16(2):845–847, June 2006.

[SFK+15] M. Schmid, J. Franck, P. Kalaria, K. A. Avramidis, G. Gantenbein, S. Illy, J. Jelonnek, I. Gr. Pagonakis, T. Rzesnicki, and M. Thumm. Gyrotron development at KIT: FULGOR test facility and gyrotron concepts for DEMO. *Fusion Engineering and Design*, 2015. 28th Symposium of Fusion Technology (SOFT 28).

[Shv14] G. Shvets. Plenary talk: Applications of electromagnetic metamaterials to vacuum electronics devices and advanced accelerators. In *15th IEEE International Vacuum Electronics Conference (IVEC)*, April 2014.

[SI08] S. Sabchevski and T. Idehara. Resonant cavities for frequency tunable gyrotrons. *International Journal of Infrared and Millimeter Waves*, 29(1):1–22, 2008.

[SJB04] K. Sing, P. K. Jain, and B. N. N. Basu. Analysis of a coaxial waveguide corrugated with wedge-shaped radial vanes considering azimuthal harmonic effects. *Progress in Electromagnetics Research*, 47:297–312, 2004.

[SJO+90] R. R. Stone, R. A. Jong, T. J. Orzechowski, E. T. Scharlemann, A. L. Throop, B. Kulke, K. I. Thomassen, and B. W. Stallard.

An FEL-based microwave system for fusion. *Journal of Fusion Energy*, 9(1):77–101, 1990.

[SKK+09] K. Sakamoto, A. Kasugai, K. Kajiwara, K. Takahashi, Y. Oda, K. Hayashi, and N. Kobayashi. Progress of high power 170 GHz gyrotron in JAEA. *Nuclear Fusion*, 49(9):095019, 2009.

[SKK+13] U. Singh, N. Kumar, A. Kumar, V. Yadav, and A. K. Sinha. Towards the design of 100 kW, 95 GHz gyrotron for active denial system application. In *14th IEEE International Vacuum Electronics Conference (IVEC)*, May 2013.

[SKO+13] K. Sakamoto, K. Kajiwara, Y. Oda, K. Hayashi, K. Takahashi, T. Kobayashi, and S. Moriyama. Status of high power gyrotron development in JAEA (session 6A). In *14th IEEE International Vacuum Electronics Conference (IVEC)*, May 2013.

[SKT+97] K. Sakamoto, A. Kasugai, M. Tsuneoka, K. Takahashi, Y. Ikeda, T. Imai, T. Nagashima, M. Ohta, T. Kariya, K. Hayashi, Y. Mitsunaka, Y. Hirata, Y. Itoh, and Y. Okazaki. Development of 170 GHz/500 kW gyrotron. *International Journal of Infrared and Millimeter Waves*, 18(9):1637–1654, 1997.

[Smy68] William R. Smythe. *Static and dynamic electricity*. International series in pure and applied physics. McGraw-Hill, New York, 3rd edition, 1968.

[Sta05] Weston M. Stacey. *Fusion Plasma Physics*. Physics textbook. Wiley-VCH, Weinheim, 2005.

[Sta10] Weston M. Stacey. *Fusion*. Wiley-VCH, Weinheim, 2nd edition, 2010.

[TAA+07] M. Thumm, S. Alberti, A. Arnold, P. Brand, H. Braune, G. Dammertz, V. Erckmann, G. Gantenbein, E. Giguet, R. Heidinger, J.-P. Hogge, S. Illy, W. Kasparek, H. P. Laqua,

F. Legrand, W. Leonhardt, C. Lievin, G. Michel, G. Neffe, B. Piosczyk, M. Schmid, K. Schworer, and M. Q. Tran. EU megawatt-class 140-GHz CW gyrotron. *IEEE Transactions on Plasma Science*, 35(2):143–153, April 2007.

[TAB+01] M. Thumm, A. Arnold, E. Borie, O. Braz, G. Dammertz, O. Dumbrajs, K. Koppenburg, M. Kuntze, G. Michel, and B. Piosczyk. Frequency step-tunable (114-170 GHz) megawatt gyrotrons for plasma physics applications. *Fusion Engineering and Design*, 53(1-4):407 – 421, 2001.

[TBP13] C. H. Townes, N. G. Basov, and A. M. Prochorov. Physics 1964 (Nobel Lecture: Production of coherent radiation by atoms and molecules). In *Physics, 1963–1970*, pages 53 – 119. Elsevier, 2013.

[TFK+15] M. Thumm, J. Franck, P. C. Kalaria, K. A. Avramidis, G. Gantenbein, S. Illy, I. G. Pagonakis, M. Schmid, C. Wu, J. Zhang, and J. Jelonnek. Towards a 0.24-THz, 1-to-2-MW-class gyrotron for DEMO. *Terahertz Science and Technology*, 8(3):85–100, 2015.

[Thu16] Manfred K. A. Thumm. *State-of-the-Art of High Power Gyro-Devices and Free Electron Masers (Update 2015)*. KIT Scientific Reports 7717. KIT Scientific Publishing, Karlsruhe, 2016.

[TLSJ15] S. Takayama, G. Link, M. Sato, and J. Jelonnek. Possibility for iron production using high-power millimeter waves. *IEEE Transactions on Plasma Science*, 43(10):3517–3521, Oct 2015.

[Tos13] Silvano Tosti, editor. *Tritium in fusion: production, uses and environmental impact*. Chemistry research and applications. Nova Science Publ., New York, NY, 2013.

[Tsi93] Sh. E. Tsimring. Limiting current of helical electron beams in gyrotrons. *International Journal of Infrared and Millimeter Waves*, 14(4):817–840, 1993.

[Vaj69] Lev Albertovič Vajnštejn. *Open resonators and open waveguides*. Golem Press, Boulder, Colo., 1969.

[VO74] S. N. Vlasov and I. M. Orlova. Quasioptical transformer which transforms the waves in a waveguide having a circular cross section into a highly directional wave beam. *Radiophysics and Quantum Electronics*, 17(1):115–119, 1974.

[VZO+69] S. N. Vlasov, G. M. Zhislin, I. M. Orlova, M. I. Petelin, and G. G. Rogacheva. Irregular waveguides as open resonators. *Radiophysics and Quantum Electronics*, 12(8):972–978, 1969.

[VZO76] S. N. Vlasov, L. I. Zagryadskaya, and I. M. Orlova. Open coaxial resonators for gyrotrons. *Radio Engineering and Electronic Physics*, 21:96–102, 1976.

[WATJ15] C. Wu, K. A. Avramidis, M. Thumm, and J. Jelonnek. An improved broadband boundary condition for the RF field in gyrotron interaction modeling. *IEEE Transactions on Microwave Theory and Techniques*, 63(8):2459–2467, Aug 2015.

[WGL+08] D. Wagner, G. Grünwald, F. Leuterer, A. Manini, F. Monaco, M. Münich, H. Schütz, J. Stober, H. Zohm, T. Franke, M. Thumm, G. Gantenbein, R. Heidinger, A. Meier, W. Kasparek, C. Lechte, A. Litvak, G.G. Denisov, A.V. Chirkov, E.M. Tai, L.G. Popov, V.O. Nichiporenko, V.E. Myasnikov, E.A. Solyanova, S.A. Malygin, F. Meo, and P. Woskov. Status of the new multi-frequency ECRH system for ASDEX Upgrade. *Nuclear Fusion*, 48(5):054006, 2008.

[WPI+16a] C. Wu, I. Pagonakis, S. Illy, M. Thumm, G. Gantenbein, and J. Jelonnek. Preliminary studies on multistage depressed collectors for fusion gyrotrons. In *German Microwave Conference (GeMiC)*, pages 365–368, March 2016.

[WPI+16b] C. Wu, I. G. Pagonakis, S. Illy, G. Gantenbein, M. Thumm, and J. Jelonnek. 3D simulation of a realistic multistage depressed collector for high-power fusion gyrotrons. In *17th IEEE International Vacuum Electronics Conference (IVEC)*, April 2016.

[Wu08] Y. Wu. Conceptual design of the China fusion power plant FDS-II. *Fusion Engineering and Design*, 83(10-12):1683 – 1689, 2008. 8th International Symposium on Fusion Nuclear Technology ISFNT-8 SI.

[Wu16] Chuanren Wu. Private communication, 2016.

[Zap12] V. E. Zapevalov. Evolution of the gyrotrons. *Radiophysics and Quantum Electronics*, 54(8-9):507–518, 2012.

[ZBD+04] V. E. Zapevalov, A. A. Bogdashov, G. G. Denisov, A. N. Kuftin, V. K. Lygin, M. A. Moiseev, and A. V. Chirkov. Development of a prototype of a 1-MW 105-156-GHz multifrequency gyrotron. *Radiophysics and Quantum Electronics*, 47(5-6):396–404, 2004.

[ZGS04] G. I. Zaginaylov, Y. V. Gandel, and S. A. Steshenko. Singular integral equation approach in the theory of coaxial cavity gyrotrons. In *5th International Kharkov Symposium on Physics and Engineering of Microwaves, Millimeter, and Submillimeter Waves*, volume 2, pages 483–485, June 2004.

[Zha16] Jianghua Zhang. Influence of emitter surface roughness and emission inhomogeneity on efficiency and stability of high power fusion gyrotrons, 2016.

[ZIP+16] J. Zhang, S. Illy, I. Gr. Pagonakis, K. A. Avramidis, M. Thumm, and J. Jelonnek. Influence of emitter surface roughness on high power fusion gyrotron operation. *Nuclear Fusion*, 56(2):026002, 2016.

[ZM11] G. I. Zaginaylov and I. V. Mitina. Electromagnetic analysis of coaxial gyrotron cavity with the inner conductor having corrugations of an arbitrary shape. *Progress in Electromagnetics Research B*, 31:339–356, 2011.

[ZMPT84] V. E. Zapevalov, S. A. Malygin, V. G. Pavel'ev, and Sh. E. Tsimring. Coupled-resonator gyrotrons with mode conversion. *Radiophysics and Quantum Electronics*, 27(9):846–852, 1984.

[ZT90] V. E. Zapevalov and Sh. E. Tsimring. Multibeam gyrotrons. *Radiophysics and Quantum Electronics*, 33(11):954–960, 1990.

[ZT00] S. Zhang and M. Thumm. Eigenvalue equations and numerical analysis of a coaxial cavity with misaligned inner rod. *IEEE Transactions on Microwave Theory and Techniques*, 48(1):8–14, Jan. 2000.

[ZT05] H. Zohm and M. Thumm. On the use of step-tuneable gyrotrons in ITER. *Journal of Physics: Conference Series*, 25(1):274–282, 2005.

Acknowledgement

First and foremost I would like to thank Prof. Dr. John Jelonnek for giving me the opportunity to contribute to the very exciting research on high-power gyrotrons and for his steady support at the Institute for Pulsed Power and Microwave Technology (IHM) at KIT. Sincere thanks also to Prof. Dr. Olgierd Dumbrajs for volunteering as Co-Referee of this thesis and for our fruitful collaboration during the past years.

I am furthermore grateful to Prof. Dr. Manfred Thumm, Prof. Dr. Edith Borie, Dr. Stefan Illy, Dr. Gerd Gantenbein, Dr. Kostas Avramidis, Dr. Ioannis Pagonakis, Dr. Zisis Ioannidis, Martin Schmid, Dr. Bernhard Piosczyk, Dr. Jianbo Jin, Dr. Tomasz Rzesnicki, Dr. Giovanni Grossetti and Dr. Sergey Soldatov for your continuous support and counsel during my PhD work.

Many thanks to Dr. Andreas Schlaich, Dr. Amitavo Choudhury, Danilo D'Andrea, Dr. Anton Malygin, Dr. Jianghua Zhang, Parth Kalaria, Sebastian Ruess, Chuanren Wu, Vasilis Ramopoulos, Dr. Benjamin Lepers, Fabian Wilde, Martin Hochberg, Alexander Marek, and Dominik Neumaier. Without our discussions and our friendship this thesis would have been much more difficult to write.

Sincere thanks also to Thorsten Kobarg, Jörg Weggen and Maximilian Fuchs for most of the 3D figures, for technical discussions and poster print-outs; to Martina Huber, Bianka Seitz, Manuela Wettstein and Kevin Paulus for your administrative and technical support; and to the KIT Scientific Publishing team for their support and advice.

Also thanks to all other IHM colleagues, especially the lunchtime community and the skiing squad, for providing the friendly and cooperative ambience at our institute.

I am grateful for the financial support from the EUROfusion Consortium and from the Helmholtz Association. Parts of the simulations presented in this work have been carried out using the Helios supercomputer at IFERC-CSC.

Thanks also to Ray C. Robinson, Jan Vetter, Dirk Felsenheimer, Rodrigo González, Aimee A. Duffy, Marcus Füreder, Matthias Otto, Farrokh Bulsara, A. M. Yankovich, Skye Edwards, Jason L. Cheetham, Eliza S. Caird, Sven Budja, Rüdiger Brans, Sebastian Raetzel, Jae-sang Park, A. L. Ivey, Jr., Maria L. Joensen, and many of your colleagues, who contributed significantly to my motivation.

Last but not least many thanks to my family and friends. I am glad to have you in my life.

"You know, out of all the components of a fusion reactor gyrotrons are the most fascinating ones."
– Drunk TLK guy at a party

KARLSRUHER FORSCHUNGSBERICHTE AUS DEM
INSTITUT FÜR HOCHLEISTUNGSIMPULS- UND MIKROWELLENTECHNIK
(ISSN 2192-2764)

Die Bände sind unter www.ksp.kit.edu als PDF frei verfügbar oder als Druckausgabe bestellbar.

Band 1 **MATTHIAS BERINGER**
Design Studies towards a 4 MW 170 GHz Coaxial-Cavity Gyrotron. 2011
ISBN 978-3-86644-663-2

Band 2 **JENS FLAMM**
Diffraction and Scattering in Launchers of
Quasi-Optical Mode Converters for Gyrotrons. 2012
ISBN 978-3-86644-822-3

Band 3 **MATTIA DEL GIACCO**
Investigation of Fretting Wear of Cladding Materials in Liquid Lead. 2013
ISBN 978-3-86644-960-2

Band 4 **AMITAVO ROY CHOUDHURY**
Investigations of After Cavity Interaction in Gyrotrons
Including the Effect of Non-uniform Magnetic Field. 2013
ISBN 978-3-7315-0129-9

Band 5 **MICHAEL BETZ**
The CERN Resonant WISP Search (CROWS). 2014
ISBN 978-3-7315-0199-2

Band 6 **ANDREAS SCHLAICH**
Time-dependent spectrum analysis of high power gyrotrons. 2015
ISBN 978-3-7315-0375-0

Band 7 **DHIDIK PRASTIYANTO**
Temperature- and Time-Dependent Dielectric Measurements
and Modelling on Curing of Polymer Composites. 2016
ISBN 978-3-7315-0424-5

Band 8 **YIMING SUN**
Adaptive and Intelligent Temperature Control of Microwave
Heating Systems with Multiple Sources. 2016
ISBN 978-3-7315-0467-2

Band 9 **JIANGHUA ZHANG**
Influence of Emitter surface roughness and Emission inhomogeneity
on Efficiency and stability of high power Fusion gyrotrons. 2016
ISBN 978-3-7315-0578-5

KARLSRUHER FORSCHUNGSBERICHTE AUS DEM
INSTITUT FÜR HOCHLEISTUNGSIMPULS- UND MIKROWELLENTECHNIK
(ISSN 2192-2764)

Die Bände sind unter www.ksp.kit.edu als PDF frei verfügbar oder als Druckausgabe bestellbar.

Band 10 **ANTON MALYGIN**
Design and Experimental Investigation of a Second Harmonic 20 kW Class 28 GHz Gyrotron for Evaluation of New Emitter Technologies. 2016
ISBN 978-3-7315-0584-6

Band 11 **JOACHIM FRANCK**
Systematic Study of Key Components for a Coaxial-Cavity Gyrotron for DEMO. 2017
ISBN 978-3-7315-0652-2